D1871265

AN ELEMENTARY
MANUAL OF ZOOLOGY

AN ELEMENTARY
MANUAL OF ZOOLOGY

DESIGNED
FOR THE USE OF FOREST OFFICERS IN INDIA.
PREPARED FOR THE FOREST DEPARTMENT OF INDIA.

E. C. COTES

GYAN

Published by

Gyan Publishing House
5, Ansari Road
Daryaganj, New Delhi-110002
Phone: 011-47034999, 9811692060
E-mail: books@gyanbooks.com

Distribution Network
gyanbooks.com
India, USA, Canada, UK, Australia, France

© Publisher

All rights reserved. No part of this work may be reproduced, stored, adapted, or transmitted in any form or by any means, electronic, mechanical, photocopying, micro-filming, recording or otherwise, or translated in any language, without the prior written permission of the copyright owner and the publisher.

ISBN: 978-81-212-8768-5 (PB)
First Published, 1893

2nd Impression 2023

Printed at: Gyan Press, Delhi.

The book is sold subject to the condition that it shall not, by way of trade or otherwise, be lent, resold, hired out, or otherwise circulated without the prior publisher's written consent.

AN ELEMENTARY MANUAL OF ZOOLOGY
Author: E. C. Cotes

AN ELEMENTARY

MANUAL OF ZOOLOGY

DESIGNED

FOR THE USE OF FOREST OFFICERS IN INDIA.

PREPARED FOR THE FOREST DEPARTMENT OF INDIA.

BY

E. C. COTES,

DEPUTY SUPERINTENDENT OF THE INDIAN MUSEUM, CALCUTTA, AND LECTURER ON ZOOLOGY AT THE IMPERIAL FOREST SCHOOL, DEHRA DUN.

CALCUTTA:

OFFICE OF THE SUPERINTENDENT, GOVERNMENT PRINTING, INDIA.

1893.

PREFACE.

THE following work has been drawn up for the Forest Department of the Government of India.

It is designed to meet a want which has been felt in the Imperial Forest School, Dehra Dun, for a short elementary manual to cover the zoological teaching of that institution.

Slight as it is, the difficulties which have had to be encountered in its preparation will be readily appreciated by those who have attempted to find their way through the fragmentary literature upon which dependence has still to be placed in studying many of the groups of Indian animals. The Vertebrata of India, it is true, have been systematically described; indeed the series of volumes which are being prepared under the able editorship of Mr. W. T. Blanford will leave, when completed, little to be desired in this section. Unfortunately the Vertebrata form numerically but a small group when compared with the myriad hosts of insects and other lowly creatures which make up the bulk of the animal species occurring in India. In this larger field students have been few and far between, and although much excellent work of a more or less isolated nature has been done, the subject is so vast and the observations generally have been so little systematized that, except in a few stray orders and families, they have hitherto been almost entirely unavailable for teaching purposes. None of the general text-books of the day again deal in sufficient detail with the particular animals with which the Indian Forest Officer is concerned, to be anything like complete guides in themselves, though they are invaluable for purposes of reference. In putting together his notes therefore, the writer has had to cull information from very numerous and widely different sources. He has also been hampered to some extent by the necessity of not overstepping the bounds of what can be gone through in the very limited time which the Dehra Dun students are able to devote to the subject. It is hoped, however, that the result will be sufficient to enable the young Forest Officer to obtain, without undue labour, a sound elementary knowledge of the classification and relative importance of the commoner and more noticeable animals which come under his observation, with some general idea of their various habits.

In the chapters devoted to the Insecta the writer has been able to draw upon the experience gained during the nine years he has been in charge of the Entomological Section of the Indian Museum in Calcutta. With regard to the remaining groups no claim is put forward to originality of treatment, though it may be stated that the information has been brought together in the course of actual teaching and that the practical work is in each case described afresh from common Indian species, which occasionally differ in matters of detail from the forms dealt with in existing text-books.

Since no reliable knowledge of Zoology can possibly be imparted by mere theoretical teaching a very considerable portion of the manual will be found to be devoted to directions for the dissection and examination of actual specimens. The dissections described can easily be performed in class by the students themselves, with the help of the extremely simple appliances that are available, while the specimens required are only such as are readily procurable in India. The classification is chiefly based upon structural characters which the students are intended, wherever practicable, to confirm for themselves by actual examination of specimens.

Attention has been especially directed to the classification of the groups which contain animals that are of practical importance in Indian forests, the others being merely alluded to shortly, in cases where it is necessary to mention them at all, in order to avoid giving an erroneous idea of the nature of the series. For instance, the classification of the snakes is dealt with at some length, as it is important that the students should be able to distinguish harmless species from poisonous ones, whereas large groups, like that of the Mollusca, are passed over in a few words as comparatively unimportant.

Information of a practical nature, connected with what has been done in various parts of the world to mitigate damage by insects related to those which are destructive in Indian forests, has been added in cases where it seemed likely to be useful; but the subject is one which has as yet been so little taken up in India that it has been thought safest not to attempt to go into it at any length.

At the end of the manual a brief note has been inserted to give the necessary instructions for preparing zoological specimens for preservation in the School Museum, where a representative series, illustrative of the chief groups of Indian forest animals of all kinds, is rapidly accumulating.

In working over the course with the students it will be necessary for the lecturer to link together the various portions of the manual by some slight explanation of the meaning of the system of classification that is indicated. And at least a few words should be added upon the subject of the fundamental theories which form the basis of the science of modern Biology. Little or no reference has been made to this extensive theme in the manual itself, for it has been thought best not to burden a practical work of the kind which is intended purely for local teaching, with theoretical matter, which is already set forth in the general zoological text-books of the day in a manner that would make any attempt upon the writer's part to deal with it afresh appear presumptuous. The school library possesses a copy of Sedgwick's edition of Claus' *Text-book of Zoology*, the introduction to which comprises an admirable treatise upon the subject, and this work is recommended for the use of students who desire further guidance.

In describing the dissections which form part of the course, the writer has constantly referred to the more detailed descriptions of allied types as given in Huxley and Martin's *Practical Biology*, Marshall and Hurst's *Practical Zoology*, and Parker's *Zootomy*, and he has generally followed the nomenclature adopted in these admirable works. In the general zoological portion of the manual the more important works consulted have been, Sedgwick's edition of Claus' *Text-book of Zoology*, Hatchett Jackson's edition of Rolleston's *Forms of Animal Life*, Huxley's *Anatomy of Invertebrated Animals*, and the volumes by Day, Boulenger, Oates and Blanford[1] in the series of the *Fauna of British India*, also Jerdon's *Birds of India* and Oates' *Birds of British Burma*. Information of a practical nature has also been taken from Clifford's *Notes on Forest Zoology*, and from what has been given verbally to the writer by Indian Forest Officers. In the chapters on Entomology constant reference has been made to the papers which have appeared in the periodical *Indian Museum Notes*, and to the masterly publications of the United States Entomological Department. Amongst numerous other sources of information special mention may be made of Packard's *Guide to the Study of Insects*, Comstock's *Introduction to Entomology*, Thompson's report on *Insects destructive to Woods and Forests*, and papers by numerous writers which have appeared in the *Indian Forester*.

Acknowledgment is also due to Messrs. Fernandez, Bagshawe, and Gamble for the interest they have taken in the Zoology teaching, and for the cordial help they have afforded in connection with the arrangements necessary for the organisation of practical work in class.

CALCUTTA: E. C. C.
The 26th May 1893.

[1] The first volume of Hampson's work in the same series was received after the chapter upon the Lepidoptera had been written.

MANUAL OF ZOOLOGY.

CONTENTS.

B

SYSTEMATIC COURSE.

GENERAL CLASSIFICATION OF THE CHIEF GROUPS OF THE ANIMAL KINGDOM.

I. Protozoa.—Animals each consisting of a single cell.
E.g., Amœba, Paramecium, Vorticella, Opalina.

II. Metazoa.—Animals each consisting of a number of cells.

A. Cœlenterata.—Multicellular animals without body cavity.
E.g., Hydra, sea anemone, coral polyp, sponge.

B. Cœlomata.—Multicellular animals provided with a body cavity.

Radiate animals. (1) **Echinodermata.**—Cœlomate animals which when adult are radially symmetrical, their bodies being arranged pentagonally and supported by a calcareous armour.
E.g., Star-fishes, sea-urchins, sea-lilies, sea-cucumbers.

trical animals. — Elongated animals without ventral foot, jointed limbs or notochord.

(2) **Vermes** (worms).—An ill-defined group of elongated bilaterally symmetrical Cœlomata, with paired excretory canals, but without jointed appendages, ventral foot, or notochord, include a large number of very different groups, of which the following may be noticed :—

Unsegmented worms.

Body flat. (*a*) **Platyhelminthes.**—Flattened worms without true segmentation.
E.g., Liver-flukes, tape-worms, and free living sea forms, such as the gliding worm.

Body round. (*b*) **Nemathelminthes** (Thread-worms).—Round or thread-like unsegmented worms, such as the Nematode and Acanthocephalous worms parasitic on animals and plants.

Segmented worms.

With blood-vessels. (*c*) **Annelida.**—Segmented worms, with definite blood-vessels.
E.g., Earthworm, leech, and numerous sea-worms.

Without blood-vessels. (*d*) **Rotatoria.**—Imperfectly segmented worms without definite blood-vessels, but with characteristic retractile ciliated apparatus at the mouth.
E.g., The wheel animalcule of stagnant pools.

Bilaterally symme

With peculiar ventral foot.

Without distinct head. / With distinct head.

(3) **Mollusca.**—Bilaterally symmetrical, unsegmented, multicellular animals, with peculiar ventral foot, and in many cases a calcareous shell.

(*a*) **Lamellibranchiata.**—Molluscs without distinct head, but with bilobed mantle, which secretes a bivalve shell.
E.g., Oyster, mussel, and scallop.

Without arms. (*b*) **Gastropoda.**—Molluscs with distinct head which may bear tentacles but not arms, with undivided mantle, which usually secretes a dorsal shell.
E.g., Snails and slugs.

With arms. (*c*) **Cephalopoda.**—Molluscs with distinct head and a number of sucker-bearing arms round the mouth.
E.g., Nautilus, ammonite, cuttle-fish.

With jointed limbs covered with armour.

(4) **Arthropoda.**—Bilateral segmented Cœlomata with jointed limbs; the muscles supported by armour plating.

Water breathes. (*a*) **Crustacea.**—Breathing water by means of branchiæ.
E.g., Lobster, crab, prawn, shrimp.

Air breathes.

(*b*) **Myriapoda.**—Air-breathing Arthropods, with more than four pairs of jointed legs.
E.g., Centipede and millepede.

(*c*) **Arachnida.**—Air-breathing Arthropods with four pairs of jointed legs.
E.g., Spiders, mites, ticks, scorpions.

(*d*) **Insecta.**—Air-breathing Arthropods, with three pairs of jointed legs. Include all insects.

With notochord.

(5) **Chordata.**—Bilaterally symmetrical animals, the body supported by an internal median rod (notochord).

(*a*) **Urochorda.**—The notochord present only in the larval stage.
E.g., Ascidian.

(*b*) **Cephalochorda.**—The notochord persistent throughout life, without brain or lower jaw.
E.g., Amphixous.

(*c*) **Cyclostomi.**—The notochord persistent throughout life, with brain but without lower jaw.
E.g., Lamprey.

(*d*) **Vertebrata.**—The notochord modified into a vertebral column (back-bone). Brain and lower jaw both present.
Include Fish, Amphibia, Reptiles, Birds, and Mammals.

I.—PROTOZOA.

The chief point to remember about the Protozoa is that they are animals which each consist of only one cell. Each Protozoon therefore may be appropriately compared to one of the vast army of cells, of which the tissues of all the higher animals are built up. In its simplest form a Protozoon consists of a little speck of transparent jelly (protoplasm) with a nucleus of denser matter. This simple structure is often obscured by shells of chalk or flint and other matters secreted by the jelly. The students will be able to examine a few of the common forms of Protozoa for themselves, through the microscope (see practical work). The following may be noticed:—

(1) *Amœba.*—This is a minute speck of jelly-like protoplasm which crawls along in stagnant water. Its movements are effected by the protrusion and retraction of little finger-like processes of the jelly which are known as pseudopodia. It feeds by encircling food particles, which it meets with on the water, with its finger-like pseudopodia, thus ingulfing them into the jelly. After absorbing the nutriment out of the particle, the shells and other waste portions are excreted by simply flowing away and leaving them behind. The Amœba reproduces itself by splitting into two parts, each of which grows to be an animal like the parent.

(2) *Paramecium.*—This is a minute slipper-shaped animalcule, which swims rapidly about in stagnant water by means of little hair-like processes known as cilia. These cilia also serve to sweep food particles to one particular spot in the body which serves as a mouth. Reproduction, which is often preceded by the temporary joining up of two individuals, is brought about much as in the case of the Amœba. (See practical work.)

(3) *Vorticella.*—This is a little transparent bell-shaped animalcule common in stagnant water. It usually has a contractile stalk, with which it anchors itself to weeds. Like Paramecium, it is provided with numerous hair-like cilia, which it uses much in the same way. It reproduces itself by simply splitting into two, also by the joining up of a small free swimming individual with a larger attached individual, followed by a period of rest and ultimate division into numerous minute spores, each of which eventually developes into an animal like the parent. (See practical work.)

Besides such simple types as *Amœba, Paramecium,* and *Vorticella,* we may notice *Noctiluca,* which has luminous protoplasm and sometimes rises in such numbers to the surface of the sea as to give a phosphorescent appearance to the water, and *Globigerina*, which secretes minute calcareous shells found in vast numbers at the bottom of some parts of the ocean, where they constitute a fine mud-like substance known as Ooze, also *Nummulites*, whose shells are largely found in limestone rocks (see specimens), and Radiolaria with minute flint shells, which are often most wonderfully sculptured. An easily procured and convenient Protozoon for the students to examine is the *Opalina* parasitic on the frog (see practical work).

II.—METAZOA.

A.—CŒLENTERATA.

The Cœlenterata comprise all the jelly-fish, sea-anemones, coral polyps and sponges, the commonest fresh-water representative being *Hydra*, which is sometimes to be found in tanks in India. The Cœlenterata differ from the Protozoa in having the body made upon a number of cells. The most important point to notice is that, in its essential features, the body is little more than a bag. Food is taken into the interior of the bag, which thus becomes the digestive tract. The cells of which the body is composed each absorbs its nutriment direct, so there is no need for blood-vessels or for any space between the digestive tract and the outer layer of the body. The Cœlenterata thus differ from all the groups of higher animals (Cœlomata), which have, at one time or other of their existence, a well-marked space (body cavity) in which the digestive tract is suspended.

Coral is the calcareous matter which is secreted by colonies of coral polyps, each of which is extremely simple in structure. The complicated structure presented by the coral itself is chiefly due to the process of budding, by which the original animals surround themselves with out-growths, each of which becomes a perfect polyp and secretes coral, though it remains attached to the original stem. In like manner the bath-room sponge is the fibrous skeleton secreted by a colony of sponge polyps, each of which is of extremely simple structure, though it differs in some particulars from the coral polyp. The students should examine and sketch pieces of coral and sponge so as to understand the process by which these substances have been built up by the polyps.

A typical instance of a Cœlenterate animal is Hydra. This is a little jelly-like creature which is to be found in tanks. It has a bag-shaped body, with a series of hollow tentacles around the mouth. It is usually to be found attached to water plants by its basal portion. The body wall is composed of two layers of cells. The inner layer consists of large cells, many of them furnished with cilia, which serve to propel the contents of the digestive cavity. These cells also contain chlorophyll particles. The outer layer consists of small cells, amongst which are a number of curious explosive cells (thread cells) which assist in seizing and paralysing prey. Reproduction takes place in two ways. One of these is by the formation of out-growths or buds, which develop into animals like the parent and drop off. The other is by the production of two different kinds of minute cells, known as Ova and Spermatozoa respectively. These specialised cells are merely out-growths of the cells of the body wall. The spermatozoan cell is free swimming. It joins up with an ovum cell, which afterwards drops off and develops into a Hydra like the parent.

B.—CŒLOMATA.

(1)—ECHINODERMATA.

The Echinodermata are a group of animals which are entirely confined to the sea. They are therefore of little importance from an Indian Forester's point of view, and the only reason for noticing them is on account of their being so well known and distinct. Like other Cœlomata, they are furnished with a body cavity around the digestive tract. When full grown they are radially symmetrical, their bodies being usually arranged pentagonally and supported by calcareous armour. They have a more or less developed system of nerves around the mouth, also blood-vessels, and a very characteristic system of vessels containing water (water vascular system), which subserves locomotion. The chief groups are the star-fishes (Asteroidea), the sea-urchins (Echinoidea), sea-lilies (Crinoidea), and the sea-cucumbers (Holothuroidea). The students should examine the fossil Echinoderms in the School Museum, and make out such features of external anatomy as the mouth, anus and radiate arrangement of the body, but no detailed acquaintance with the group will be required.

(2)—VERMES (WORMS).

This is an ill-defined group of Cœlomate animals. It comprises a large number of very distinct creatures which differ from each other so widely that some zoologists prefer to class the different groups altogether separately from each other. The most characteristic features which the various representatives have in common are their elongated shape, their bilateral symmetry and their paired excretory canals. Even these very general characteristics, however, are so subject to exception, that perhaps the best way to characterise the group is to compare it with the other large sections of the Cœlomata and show how the Vermes differ in each case. Thus from the Echinodermata they differ in being bilaterally symmetrical throughout life; from the Mollusca they differ in not possessing the characteristic ventral foot of that group; from the Arthropoda they differ in not possessing jointed armour-clad appendages; while from the Chordata they differ in not possessing a notochord.

For an account of the structure of a typical representative of the group, see the chapter on the dissection of the earthworm. Worms are divided into a large number of sections, but the following are the only ones that the students need notice:—

(*a*) **Platyhelminthes.**—These are flattened worms without true segmentation. They include a large number of very different animals, of which we may notice:—

(1) The liver-flukes, which in their adult form are parasitic in the bodies of sheep and other animals, while in their immature form they pass through a series of complicated transformations in the tissues of snails, etc.

(2) Tape-worms, which in their adult condition are parasitic in the digestive organs of mammals, while in their immature form they inhabit the tissues of animals of many different kinds which the eventual host eats. They usually present a specialised form of spurious segmentation due to the multiplication of the posterior part of the body in which the reproductive organs are situated.

(3) Free living sea forms, such as the gliding worms (Turbellaria), which are interesting as comprising some of the simplest organisms of the group.

(*b*) **Nemathelminthes.**—These are round or thread-like unsegmented worms, comprising a large number of forms parasitic on animals and plants. The two most important groups are the Nematoda and the Acanthocephala.

The Nematode worms include the parasitic round and pin worms common in the digestive tracts of mammals, the parasite (Dochmius)[1] which passes through one stage of its existence in the human digestive tract and another stage in the earth, minute Filaria, which often infest the blood of mammals and birds in India, and which have been supposed to pass through a stage of their existence in the mosquito, minute Trichinæ, which make their way into the muscles of mammals, producing a disease known as Trichinosis, also thread-worms, which are parasitic on locusts and other insects, and numerous species which are parasitic on plants. Nematode worms for the students to examine are usually to be found in large numbers in the tissues of the Dehra fresh-water crab—*vide* chapter on the dissection of this creature.

[1] Said to produce the so-called *kala-azar* and *beri-beri* diseases which have caused an enormous number of deaths of late years in Assam.

The Acanthocephala are parasitic worms armed with a protrusile proboscis furnished with hooks. They chiefly inhabit the digestive tracts of Vertebrate animals, sometimes passing through a stage of their existence in the grubs of insects (*e.g.*, cockchafers) upon which their ultimate host feeds.

(*c*) **Annelida.**—These are worms which have their bodies divided internally into well-marked segments and are provided with definite blood-vessels. They comprise a number of groups of both sea and land worms, but the only ones that need here be noticed are the earth-worms and leeches which the students will be able to examine for themselves. Earth-worms are of great importance as soil-changers, for they are continually bringing up soil from below and depositing it above ground. Leeches, some of which are common forest pests in India, pass their early stages amongst vegetation or in water, where they are said to feed upon molluscs and other small animals. It is only in their later stages of growth that they affix themselves to mammals and fill themselves with warm blood.

(*d*) **Rotatoria.**—These are imperfectly segmented worms without definite blood-vessels. The commonest representative is the little transparent wheel animalcule to be found in stagnant pools about Dehra. It may be recognised by the curious ciliated apparatus in front of the mouth, which serves to capture food particles in the water.

(3)—Mollusca.

The Mollusca comprise all such animals as snails, slugs; oysters and cuttlefish. They are bilaterally symmetrical, unsegmented, cœlomate animals, with peculiar ventral foot and in many cases a calcareous shell. For an account of the structure of a typical representative of the Mollusca, see the chapter on the dissection of the snail. The most important groups are the Lamellibranchiata, the Gastropoda, and the Cephalopoda.

(*a*) **Lamellibranchiata.**—These are Molluscs which have no distinct head differentiated from the rest of the body. They have a bilobed fold of integument (mantle), which secretes a bivalve shell. They include a vast number of water creatures, such as oysters and mussels, but are of no very great importance from a forester's point of view. The students should examine the bivalve shells in the Museum and sketch one or two typical specimens.

(*b*) **Gastropoda.**—These are Molluscs with a distinct head which sometimes bears tentacles but never arms. They have an undivided mantle which usually secretes a dorsal shell. They include all the land and water snails which have a shell developed, and the slugs where the shell is rudimentary. For details of structure, see the chapter on the dissection of the snail. The Museum contains numerous Gastropoda shells for the students to examine.

(*c*) **Cephalopoda.**—These are sea creatures, with distinct head and a number of sucker-bearing arms around the mouth. They include all the Cuttle-fishes, Squids, and Belemnites which are without external shells, and the Nautilus and Ammonite, where a specialised many-chambered shell is developed. The group is of very little importance from a forester's point of view, but the students should notice the Nautilus shells and the Ammonite and Belemnite fossils in the Museum and sketch a typical representative of each.

(4)—Arthropoda.

The Arthropoda are a vast group of bilaterally segmented Cœlomata, which are characterised by the possession of jointed appendages supported by external armour. For accounts of the structure of typical representatives, see the chapters on the dissection of the fresh-water crab, the cockroach and the mulberry silk-worm. The chief groups comprised by the Arthropoda are the Crustacea, the Myriapoda, the Arachnida, and the Insecta.

(*a*) **Crustacea.**—The Crustacea comprise a large assortment of aquatic Arthropoda, which breathe by means of branchiæ. They include lobsters, crayfish, shrimps, prawns, hermit crabs, shore crabs, fresh-water crabs, king crabs, barnacles, water-fleas, and a number of less well-known forms. From the forester's point of view they are of no very great importance, and the students will learn all that is requisite about their structure from the dissection of the fresh-water crab. Attention should also be given to the representatives of the group in the School Museum, sketches being made of a few typical forms.

(*b*) **Myriapoda.**—The Myriapoda are a small group of air-breathing Arthropoda, which are characterised by having a large series of legs on either side of the body. They are divided into the Centipedes (Chilopoda) and the Millipedes (Chilognatha). The Centipedes have but one pair of legs attached to each segment of the body and are carnivorous in their habits. They are furnished with a large pair of poison claws, which are modifications of the second pair of thoracic appendages. The Millipedes have two pairs of legs attached to most of the segments of the body. They feed chiefly on vegetable matter. They have no poison claws and are quite harmless. The students should examine and sketch a Centipede and a Millipede from the Museum collection and make out the structure of the legs and mouth parts in each case.

(*c*) Arachnida.—The Arachnida are a large group of air-breathing Arthropoda, which are characterised by the possession of four pairs of jointed legs. The only orders that need be noticed are the Acarina, the Araneida and the Scorpionidea, but it should be remembered that there are other orders which are equally distinct from each other (*e.g.*, the Book-scorpions, the Pedipalpi and the Solfugæ, some of which are represented in the collections of the School Museum).

The *Acarina* (Mites) are mostly minute creatures. They have a thick-set body and unsegmented abdomen fused to the thorax. Their mouth parts are formed chiefly for biting and sucking. Many of the mites are parasitic on plants and animals, *e.g.*, the so-called "red spider," which attacks tea-bushes, and the itch mite of man. The ticks, which are common forest pests, are merely large mites. They are characterised by their strong dorsal shield and large protrusile toothed jaws. They usually pass their early stages amongst vegetation, whence they crawl on to mammals and birds to suck blood. The efficacy of flowers of sulphur, either dusted on to the plants or sprayed in a mixture of soap and water, for destroying "red spider" and other mites which attack plants, has been established. This form of treatment, however, is only practicable with valuable plants occupying restricted areas, as in the case of orange trees and tea-bushes upon which it has been successfully employed. A convenient mite for students to examine and sketch is the large red velvety Trombidium, numbers of which are preserved in the School Museum.

The *Araneida* (Spiders) are characterised by their swollen unsegmented abdomen, which is separated off from the cephalothorax by a well-marked constriction. Most spiders have spinning glands situated at the posterior end of the body. The anterior pair of cephalic appendages (cheliceræ) are formidable weapons furnished with poison glands, which aid in killing the insects on which most spiders feed. The second cephalic appendage has a large, jointed palp (pedi palpus), which often has the appearance of a fifth leg. The order comprises a large number of groups of species which catch their prey in a variety of different ways. From an Indian forester's point of view the Araneida are of no very great importance. The students should examine and sketch a few typical representatives from the Museum.

The *Scorpionidea* (Scorpions) are characterised by a peculiar elongated segmented abdomen, the posterior portion of which is modified into a long tail-like appendage terminating in a large poison spine. The second pair of cephalic appendages (pedipalpi) are armed with large and powerful pincers (chelæ). Scorpions feed chiefly on spiders and insects, which they are said to catch with their chelæ and sting to death with their caudal poison spine. The sting of large individuals is said to be powerful enough to be of serious nature even to man. The students should examine and sketch a scorpion making out the features noticed above.

(*d*) **Insecta.**—The Insecta (Insects) are characterised by the possession of three pairs of jointed legs. They generally have the body divided into the three well-marked regions of head, thorax and abdomen. For accounts of the structure of typical insects, see the chapters on the dissection of the cockroach and the mulberry silk-worm. Insects comprise some hundreds of thousands of species many of which are of very great importance in Indian forests. It will therefore be necessary to consider the chief sections of the group in detail.

(*d*) INSECTA.

CLASSIFICATION OF THE MORE IMPORTANT ORDERS.

Without pupal stage.	Mouth parts for biting.		I.—Orthoptera.
	Mouth parts for sucking.		II.—Rhynchota.
With pupal stage.	Mouth parts in adult for biting.	Front wings membranous with many nervules.	III.—Neuroptera.
		Front wings horny.	IV.—Coleoptera.
		Front wings membranous with few nervules.	V.—Hymenoptera.
	Mouth parts in adult for sucking.	Hind wings rudimentary.	VI.—Diptera.
		Hind wings developed.	VII.—Lepidoptera.

I.—Orthoptera.—Pupal stage absent. Mouth parts formed for biting.
E.g., Locusts, grass-hoppers, crickets, cockroaches, mantids, stick insects, white-ants and dragon-flies.

II.—Rhynchota.—Pupal stage absent. Mouth parts converted into a jointed tube.
E.g., Plant-lice, scale insects, Cicadas, leaf-hoppers and Heteroptera.

III.—Neuroptera.—Pupal stage present. Mouth parts formed for biting. Wings with fine net-like nervules.
E.g., Ant-lions and Chrysopa flies.

IV.—Coleoptera.—Pupal stage present. Mouth parts formed for biting. Front wings converted into horny sheaths, which unite down the back in a straight line.
Include beetles of all kinds.

V.—Hymenoptera.—Pupal stage present. Mouth parts formed for biting and licking. Wings, when present, membranous with comparatively few nervules.
E.g., Bees, wasps, ants, fig-insects and most four-winged flies.

VI.—Diptera.—Pupal stage present. Mouth parts modified for piercing and sucking. Hind wings rudimentary. Front pair, when present, membranous.
Include two-winged flies of all kinds.

VII.—Lepidoptera.—Pupal stage present. Mouth parts converted into a coiled tube. Wings covered with scales.
Include butterflies and moths.

I.—Orthoptera.

The Orthoptera comprise a large number of very distinct forms. They may be defined as insects without pupal stage, with mouth parts formed for biting, the adult in most cases provided with two pairs of wings. The Orthoptera appeared very much earlier than most of the other orders of Insects in the Earth's history, and the structure that they present is of a generalised type. An account of the main features in the structure of a typical orthopterous insect is given in the chapter upon the dissection of the cockroach which the students should work over carefully.

The following is a classified list of the more important groups of Indian Orthoptera :—

I.—ORTHOPTERA GENUINA.—Front wings narrow and more or less stiff.

a. **Cursoria.**—With legs formed for running.

E.g., Blattidæ (Cockroaches).

b. **Gressoria.**—With thin weak legs suitable only for slow progression.

E.g., Mantidæ, front legs modified for catching prey.

Phasmidæ, front legs not so modified.

c. **Saltatoria.**—Hind legs modified for leaping.

E.g., Acrididæ (Locusts and grass-hoppers), with short antennæ.

Locustidæ (Tree-crickets), with long filiform antennæ; front wings set more or less vertically.

Gryllidæ (House and mole crickets), with long filiform antennæ; front wings set horizontally.

II.—ORTHOPTERA PSEUDONEUROPTERA.—Front wings as broad as the hind wings and membranous.

E.g., Termitidæ (White-ants).—Larvæ live on land.

Libellulidæ (Dragon-flies).—Larvæ live in water.

Blattidæ (Cockroaches).—These insects are common house and ship pests, but are of no great importance to the forester. They lay their eggs in horny purse-shaped capsules. The larvæ are wingless. The full-grown insects usually have wings. They pass the whole of their lives in houses and other sheltered places, where they feed on sugar, starch and refuse of all kinds. Enough will be learnt about the structure of the group in the course of dissecting the species *Periplaneta americana.*

Mantidæ.—These insects are remarkable on account of the modification of the femur and tibia of the front legs for the purpose of catching insects. The eggs are laid in masses of frothy matter which hardens so as to form a protection to the eggs. The young are much like their parents, but without wings. They feed upon similar food. Mantidæ destroy a large number of insects, some of which may be destructive ones, but they are of no great importance. The students should sketch a typical mantis from the Museum and examine the structure of the front pair of legs.

Phasmidæ.—These are curious weak creatures, which mimic dry sticks and leaves in a wonderful manner. They feed upon leaves, but are not usually numerous enough to do any appreciable damage. They are noticed chiefly as examples of the perfection to which the phenomenon of mimicry obtains amongst insects. The students should sketch a stick insect from the Museum.

Acrididæ.—This group is a most important one. It comprises not only the locusts, which are often most destructive in India, but also a vast number of grass-hoppers of all kinds, which damage crops to a greater or less extent. The Acrididæ can always be easily distinguished from other groups of leaping Orthoptera, by their short antennæ and thick-set bodies. The Acrididæ all feed upon growing plants, and the general features of their habits are somewhat similar, though they sometimes differ from each other in points that are of importance in dealing with them practically. About the most important species is the migratory locust (*Acridium peregrinum*, Oliv.) of North-Western India. In its adult stage this insect is between two and three inches in length. When it first acquires wings it is of a beautiful salmon pink, but this colour later on changes to yellow and afterwards to a dull purple. It breeds permanently in the sandy deserts of Western Rajputana and the North-west frontier, from which region flights periodically invade all parts of India. The eggs are laid in the ground, and hatch in about a month. The young larvæ are little wingless grass-hoppers, which feed on green plants of almost every kind and often do a vast amount of damage to crops. The larval stage lasts for from one to two months, during which time the insects moult their skins at intervals. When full grown they moult their skins for the last time and acquire wings. They then take flight and fly about the country in swarms, alighting at intervals to devour the crops. In the Deccan there is another migratory locust, *Acridium succinctum*, Linn., which can at once be distinguished from *Acridium peregrinum* by the sculpture on its thorax. It breeds permanently in hill ranges and belongs to a zone of more luxuriant vegetation than *Acridium peregrinum*. Its life history is very similar, but it is not found altogether amenable to the same methods of treatment. Throughout India there are a large number of other Acrididæ, which partake more or less of the migratory habits of locusts and often do a great deal of damage to crops. They are too numerous, however, to make it possible to consider them in detail in this work.

Generally speaking, it is only in seasons of drought that locusts multiply excessively. This indicates that a dry soil is favourable to them and that the extension of irrigation and forests is likely to tend in the long run to diminish their ravages.

Locusts are very subject to parasites, and the number which are devoured each year by birds is enormous, while unsuitable climatic. conditions, and specially damp, seem to be peculiarly fatal to them. *Acridium peregrinum* is known to suffer from two dipterous parasites, one of which attacks the egg and the other the mature insect. A Carabid beetle (*Colosma orientale,* Hope) devours vast numbers of the wingless larvæ, while the Rosy pastor starling (*Pastor roseus*) works almost incredible havoc amongst the flights. The Deccan locust (*Acridium succinctum*) is said to suffer from parasitic mites and Nematode worms, and there seems reason to believe that disease is also prevalent.

Locusts have been shown to be capable of being dealt with effectively in India by artificial means, but the greatest activity and energy is necessary. The old locusts can be killed when they are numb with cold in the early mornings. The eggs can be collected by hand in vast numbers; while much can be done in the way of beating the winged locusts off the crops, thus compelling them to feed on grass and other vegetation of small value. These means should all be employed, but the most effectual mode of fighting the pest is by destroying the young wingless larvæ. In the case of *Acridium peregrinum,* these can be destroyed in millions by driving them into lines of trenches or burning straw, long lines of screens three feet high, made of cloth with a narrow band of slippery waxcloth along the top, being most useful for guiding the insects into the traps. In the case of *Acridium succinctum,* which refuses to be driven in this manner, the best thing has been found to be collecting the insects by dragging bags, made of stout cloth, over the crops in such a way as to catch the young locusts.

In the School Museum is a specimen of *Acridium peregrinum,* also an illustrated diagram which gives its life history in detail. The students should examine these carefully. They should also sketch one or two Acrididæ in their note-books.

Locustidæ[1] (Tree-crickets).—These insects are much like Acrididæ, but

[1] It is unfortunate that the name Locustidæ should be applied to a group of insects other than what are popularly known as "locusts." The usage, however, is so well established that it would be difficult to change it without producing further confusion.

can easily be distinguished by their long thread-like antennæ. They are generally much more slender in general build, and have a pair of easily seen auditory organs, which are situated at the upper ends of the front tibiæ, while the corresponding organs in the Acrididæ are situated on the metathorax. Like other Saltatoria, the Locustidæ are able to produce chirping sounds. These are often so loud as to make the forest resound, especially in the rainy season. The sound is produced by scraping the bases of the front wings together. Some of the species are said to prey upon other insects, but, generally speaking, they are vegetable feeders. They lay their eggs either in the tissues of plants or in the earth. The Locustidæ have not been recorded as doing any special damage in forests, but they should be noticed on account of the well-known sounds they produce. The students should sketch one or two representative specimens from the collection. in their note-books.

Gryllidæ.—This group includes the house-crickets and the mole crickets. The Gryllidæ are closely allied to the Locustidæ and have much the same long filiform antennæ; they can be distinguished from the Locustidæ, however, by their horizontally set front wings. As in the case of the Locustidæ, many of the species produce loud chirpings. The mole-crickets have their front legs thickened and broadened for burrowing in the earth. They feed chiefly on vegetation, and have been reported in some cases as causing a good deal of damage to crops. The students should sketch one or two typical specimens from the Museum collection.

Termitidæ (White-ants).—The winged individuals of these destructive creatures differ from the corresponding form in the preceding groups in having the front pair of wings membranous, and almost entirely similar in structure to the hind wings. White-ants live together in communities, which consist of —

(1) A female or queen, with enormously distended abdomen; she is incapable of locomotion and lays all the eggs.

(2) Small ant-like neuters, which may be compared to the worker bees: they are of two kinds, *viz.*, individuals with large heads and sharp mandibles, to defend the nest, and individuals with small heads, to build the passages and nest, collect the food and do all the work of the community.

(3) Wingless larvæ, which develop into winged males and females.

(4) Winged males and females. The males and females fly out of the nest in clouds, generally after rain. Those of them that escape their numerous enemies are said to drop their wings and copulate. The female either finds her

way back to the original nest or starts a fresh nest for herself. Her abdomen gradually grows by distention of the membrane between the chitinous plates, until she becomes like a sausage, two or three inches long, with a minute head and thorax at one end. Queens in several stages of development may sometimes be found in a single nest.

White-ants attack dead wood, paper, cloth and other substances. They never willingly expose themselves to day-light but work in earthen galleries, which they build over themselves as they advance. They sometimes attack young sugarcane, also plants in nurseries, but can be removed by copious watering and by opening up the earth around the stem so as to admit light and air. They also cover trees with their earthen galleries and feed on the outer portion of the bark. When the trees are healthy, however, little or no damage is done. White-ants will not attack good teak or other really hard wood even when dry. About the most effective insecticide for use against them is kerosine oil, which is peculiarly distasteful to insect life of all kinds, but they are susceptible to almost every kind of insect poison, though the use of these substances is only practicable in some cases. It should be noticed that white-ants do good service in forests by converting fallen branches into soil. The students should examine the specimens in the School Museum.

Libellulidæ (Dragon-flies).—These insects are of no special importance, and are only noticed as they are excessively common and conspicuous. They are totally different in appearance from the other Orthoptera, their most characteristic features being their large membranous wings, supported by a fine network of veins, and their short, slender, pointed antennæ. Unlike the groups noticed above, their larvæ live in water. Dragon-flies feed on insects both in their larval and winged stages. The larva has the labium enormously developed for catching its prey, and breathes water by means of a modification of the trachææ in the posterior of the body. The students should sketch a representative specimen in the Museum. If possible they should also examine the larva.

II.—Rhynchota.

The Rhynchota comprise a large number of very distinct forms. They may be defined as insects furnished with a jointed proboscis, no pupal stage being present. The larvæ are little creatures much like their parents, but without wings. They are active throughout their entire existence. The jointed proboscis has arisen as a modification of mouth parts like those of the cockroach, but there is some doubt as to the manner in which the modification came about.[1] The proboscis is used for piercing the tissues of plants or animals and sucking up the juices upon which the Rynchota feed. Many of the species have cutaneous glands, which secrete offensive matter. These glands pour out their secretion by ducts which open on the meso and meta-thorax between the hind pair of limbs. Other species have glands which secrete wax, while others again have glands which secrete sugary matter. In the case of the wax and sugary secretion, the glands are situated on the dorsal surface of the abdomen. The following is a classified list of the more important groups of Indian Rhynchota :—

I.—APTERA.—Minute parasitic Rhynchota without wings at any stage of their existence. Include the lice, which are parasitic on birds and mammals.

II.—HOMOPTERA.—Rhynchota with front wings of the same thickness throughout.

(*a*) *Coccidæ* (Scale insects).—Tarsi one-jointed. Females usually scale-like. Male, with not more than one pair of wings. Mostly minute insects, *e.g.*, Palm scale, coffee scale, tea scale, orange scale, lac insect and cochineal insect.

(*b*) *Aphidæ* (Plant lice).—Tarsi two-jointed. Antennæ three to seven jointed. Minute insects, *e.g.*, tea aphid, spruce Chermes, American blight.

(*c*) *Psyllidæ.*—Tarsi two-jointed. Antennæ nine or ten jointed. Hind legs often thickened for jumping, *e.g.*, Mango Psylla, and Indigo Psylla.

(*d*) *Cicadidæ.*—Tarsi three-jointed. Three ocelli present. The males with sound-producing organs. Mostly large insects, *e.g.*, the common "croaker" of hill forests.

(*e*) *Fulgoridæ.*—Tarsi three-jointed. Not more than two ocelli present. Antennæ said to be inserted beneath the eyes. Include all the so-called lantern flies, many of which are brilliantly coloured and of considerable size. One of the species (*Phromnia marginella*, Oliv.) produces large quantities of a white sugary secretion, but the family is of little practical importance,

(*f*) *Cicadellidæ.*—Tarsi three-jointed. Not more than two ocelli present. Antennæ said to be inserted between the eyes,

[1] According to Professor J. B. Smith (*Science*, New York, April 1892), the whole beak consists of the fused blades of the maxillæ, the mandibles and labium being both rudimentary and the styles, contained within the beak, being merely modification of parts of the maxillæ. According to the older theory the mandibles and maxillæ form the styles, while the jointed sheath is composed of parts of the labium (probably the fused labial palps).

but this feature is difficult to make out. Include the leaf-hoppers of all kinds, *e.g.*, green-fly blight of tea, the mango Jassid, and the curious *Centrotus* which is attended by the large black (*Camponotus*) ant.

III.—Heteroptera.—Front wings have the basal portion thickened. The sub-order contains a large number of families, of which it will only be necessary to notice the following:—

(*a*) *Pentatomidæ.*—These insects can usually be recognised by their triangular shape, large scutellum and five-jointed antennæ. Include soldier-bugs, which destroy caterpillars.

(*b*) *Coreidæ.*—Can usually be recognised by the shape of the prothorax. The antennæ are four-jointed, and there is a long weak proboscis. Most of the species are yellowish in colour and have the posterior femora more or less thickened. Include the rice-sapper and other plant-feeding forms.

(*c*) *Lygæidæ.*—These are slender weak insects with four-jointed antennæ and thin proboscis. Most of the common Indian species are reddish in colour, and can be distinguished from the Coreidæ by their slender body and indented prothorax. Include the red cotton bug and other plant-feeding forms.

(*d*) *Capsidæ.*—These are small insects, which can usually be distinguished from the preceding families by the smallness or absence of the terminal (fourth) joint of the antennæ. Include the mosquito blight of tea and other plant-feeding forms.

(*e*) *Reduviidæ.*—These insects can usually be recognised by their freely movable heads, and short curved proboscis, with which the larger species can inflict a sting of some severity. They feed on insects, and probably destroy a number of injurious forms.

Coccidæ.—These minute insects do an enormous amount of damage to vegetation and are likely to be extremely destructive in Indian forests, though little has yet been recorded upon the subject. Cocoanut palms in the Laccadive Islands, also Areca palms on the Bombay Coast, and coffee bushes in Southern India and Ceylon are known to suffer to a very serious extent indeed from the attack of Coccidæ insects, while mango and orange trees, tea bushes and other plants have been reported as harbouring various species. Among useful species may be noticed the lac-insect and the cochineal. The lac-insect (*Carteria lacca*) secretes both wax and dye, the lac being poured out by glands situated on the back. The insect feeds on Dhak (*Butea frondosa*), Ber (*Zizyphus jujuba*), Pipal (*Ficus religiosa*), Kusum (*Schleichera trijuga*), Babul (*Acacia arabica*), and numerous other forest trees. It yields two crops in the year and forms a valuable article of forest produce, for India supplies the world with lac. The cochineal insect (*Coccus cacti*) secretes a valuable dye. An inferior variety flourishes on prickly pear (*Opuntia*) in India and is used to a small extent in native dyeing, but success has not hitherto attended the attempts that have been made to introduce into India the Mexican variety, which produces the best dye.

Amongst the Coccidæ the females are wingless creatures, which attach themselves to plants and gradually lose their power of locomotion, so that they often appear like little scales. The males are usually active creatures with but one pair of wings. The larvæ crawl out from beneath the mother insects, where the eggs are laid, and are liable to be carried considerable distances by the wind, thus spreading the species from tree to tree. They injure the plant by sucking up its sap by means of the delicate proboscis with which they are armed. *Remedies against scale insects.*—In America kerosine and soap emulsion applied in a very fine spray with a force pump has been found an efficient remedy. In the case of the scale insects which attack coffee and tea in India also, experiment has shown that this treatment is successful, though its cost has hitherto prevented its adoption generally. It will probably be found applicable in the case of nurseries and valuable fruit trees that happen to be attacked, but is out of the question in forests.[1] Washes made of solutions of resin soaps have also been recommended. Another system known as the "gas" treatment has been adopted upon a considerable scale for ridding orange trees in California of scale insects. It consists in generating hydrocyanic gas under a kind of movable canvas tent which is thrown over the bush. This method is said to be extremely effective, but is only likely to be applicable under very exceptional circumstances. The fact that scale insects are largely distributed by the action of the wind, which blows the young larvæ from one tree to another, makes wind breaks consisting of high trees often of value for the protection of crops which might otherwise be attacked. For trees, however, to be of any use for the purpose, it is of course essential that they should not be subject to attack from the

[1] Kerosine and soap emulsion is best made by churning with a force pump two parts of kerosine oil with one part of hot soap solution; the soap solution to be made by boiling about half a pound of soap to each gallon of water. The strength of the application must vary according to the nature of the insect to be dealt with and that of the plant attacked, but when applied in the proportion of twelve parts of water to one of the emulsion, in a very fine spray through an eddy nozzle by means of a force pump, it will kill most scale insects without appreciably injuring the plant.

species that affects the particular crop which they are intended to shelter. Information upon the subject of the trees attacked by the various species is still very imperfect in India. The School Museum possesses some specimens of lac, and any insects that can be procured should also be examined.

Aphidæ.—These minute insects do an enormous amount of damage to vegetation and are likely to prove extremely destructive in Indian forests, though little has as yet been recorded upon the subject. Apple trees in the Nilgiris and the Himalayas suffer very seriously from the attack of the species *Schizoneura lanigera*, Hausm., while tea, coffee, cinchona, mustard, spruce and bamboos in various parts of India have been reported as attacked by other species. The irritation set up by Aphidæ in feeding on the juices of plants often results in the formation of galls. This feature is specially remarkable in the case of the form (*Chermes abietis Linn.*) which attacks the twigs of the spruce fir (*Abies Smithiana*) in the North-Western Himalayas, producing a gall which may readily be mistaken for a fir cone, also in that of *Schizoneura lanigera*, Hausm., which produces a knotty growth upon the twigs and roots of apple trees. In Europe one of the best known species is *Phylloxera vastatrix* which has done an enormous amount of damage to vineyards. It has not yet been recorded from India, but is likely to occur in Kashmir.

The life history among Aphidæ is extremely complicated. The following are the main features of what usually goes on in the case of species inhabiting a temperate climate, and the same will probably prove to obtain in a more or less modified degree in India.—A single so-called " winter egg " is laid in the autumn by each female on the bark of the food plant. This hatches in the spring, producing a wingless female, which gives birth, without the intervention of a male, to other females, which may be either winged or wingless. These again reproduce themselves in a similar manner, and the process goes on through an indefinite number of generations until the weather begins to grow cold. Males and females are then produced. These copulate and the female produces, either directly or through the intervention of a further generation, the single winter egg, which, as we have seen, hatches in the following spring.

Most Aphidæ have a pair of elongated dorsal tubercles on one of the posterior segments of the abdomen. These are the ducts of simple glands which secrete the honey dew that is greedily sought after by ants. The result is that many species of Aphidæ are attended and protected by ants. The rose bushes in the Dehra school compound usually afford a plentiful supply of Aphidæ with their attendant ants for the students to examine.

Aphidæ are usually amenable to the methods of treatment noticed as applicable against Coccidæ (see page 20). Those species which do not form galls indeed can often be got rid of by very much weaker applications, though it is only in special cases that treatment of any kind is practicable.[1]

Psyllidæ.—These minute creatures are closely related to the Aphidæ, and need not be noticed in detail. Two remarkable Indian species are *Psylla cistellata,* Buckton, which causes the abortion of the shoots of the mango tree, and *Psylla isitis,* Buckton, which may be recognised by the characteristic galls it forms upon indigo plants which were reported in 1890 as suffering to an enormous extent in Bengal.

Cicadidæ.—These well-known creatures produce the continuous whirring sound which is so characteristic a feature of hill forests. The sound is produced by the male by means of specially modified lateral stigmata which can readily be seen at the base of the hind legs. The habits of the group, as observed in the temperate zone, are as follows :—The female lays her eggs in the branches of trees, where they are said to be to some extent nourished by the sap, the irritation which they set up often doing considerable injury to leading shoots. The young emerge from the eggs and crawl down to the ground, where they live through several years. They feed upon the roots of plants, but are not thought to do much injury in this stage of their existence. When at last they are full grown they emerge from the ground and transform into the winged males and females which produce the eggs. The students should examine and sketch some of the specimens in the School Museum.

Fulgoridæ.—These creatures usually spend the whole of their existence upon the food plant, where the eggs are laid and where the young wingless larvæ gradually grow into the winged adult. They are often conspicuous insects, but are of little practical importance.

Cicadellidæ.—The representatives of this group are much like the *Fulgoridæ.* The little green species *Chlorita flavescens,* Fabr., which is known in Assam as *blister blight* and in Sikkim as *green-fly,* occasionally does a good deal of damage by sucking up the juice of the tea leaf. Closely allied to it is the mango Jassid (*Idiocerus niveosparsus,* Leth.), which attacks the blossoms of the mango in a similar manner and is said to seriously interfere with the yield of

[1] In the case of the Aphis which attacks tea bushes to a small extent in Ceylon, Mr. Green writes—" A weak wash of kerosine emulsion or phenyle applied with a brush will instantly kill every Aphis. It should be applied either in the evening or on a cloudy day. Hot sunshine upon leaves freshly drenched with kerosine is apt to scorch them. One part of kerosine emulsion to eighty parts of water will make a mixture strong enough to kill every Aphis with which it comes in contact; while a still weaker mixture of phenyle, one part to two hundred and forty of water, is sufficient for the same purpose."

fruit. Spraying the trees at the time of flowering with a wash made of the arsenical insecticide known as London purple, has been found an effective remedy in the Saharunpur botanical gardens, but it is doubtful to what extent the application is practicable generally.

Heteroptera.—In habits there is little to notice about the land Heteroptera which in this respect are very similar to the Fulgoridæ and the Cicadellidæ. No very satisfactory method of dealing with them has yet been discovered, though heavy pruning combined with high cultivation is said in some cases to have proved beneficial against the mosquito blight of tea gardens. In Ceylon the cultivators are said to have a way of destroying the destructive rice-sapper by beating through the paddy with a winnow rubbed with glutinous matter to which the insects adhere.

III.—Neuroptera.

The Neuroptera have many points in common with the Orthoptera. They generally possess four membranous wings, with a thickly-set network of veins much like those of the Libellulidæ. Again they have biting mouth parts which are essentially similar in structure to the mouth parts of the cockroach. They differ from the Orthoptera in passing through a resting pupal stage. When full fed the larva spins itself up into a cocoon, where it transforms into a motionless chrysalis, from which the winged adult afterwards emerges. Two of the commonest Indian groups are the Chrysopinæ and Myrmeleoninæ.

(1) *Chrysopinæ* (Chrysopa or lace-winged flies). The imago is a slender insect, yellowish or greenish in colour, which may be recognised by its thread-like antennæ and two similar pairs of delicate transparent wings thickly beset with net-like nervules. The larvæ are active little creatures, with sharp mandibles. They feed upon Aphidæ and are believed to have a considerable effect in keeping these destructive insects in check. The eggs are set upon slender stalks about a quarter of an inch in length, little groups of them may often be seen standing on twigs where Aphidæ are numerous. The appearance they present is very much that of a number of little white pins stuck into the twig.

(2) *Myrmeleoninæ* (Ant-lions).—Amongst other places these creatures are common in the bed of the Song river near Dehra,

[1] Said to be composed of the "coagulated milk of the jak."

where the larvæ form little pits in the sand in which they lie in wait for passing ants and other insects. The imago is a large-winged insect very much like a dragon-fly, from which, however, it can at once be distinguished by its many-jointed, club-shaped, antennæ.

IV.—Coleoptera.

This group comprises all the beetles. These insects can be recognised by the shape of the horny front wings (Elytra), which meet down the back in a straight line. The mouth parts are essentially similar in structure to those of the cockroach. The prothorax is freely movable upon the mesothorax, and there is a well-marked pupal stage. Some ninety thousand different species of beetles have been described, and of these a very considerable proportion occur in India. They present a large number of well-marked groups, in many cases with very different habits. In this manual it will only be necessary to notice a few of the more important families. Beetles are classified, according to the structure of their tarsi, into the following groups :—

I.—**Pentamera.**—All three pairs of legs with five-jointed tarsi. The most important families are the *Cicindelidæ, Carabidæ, Dermestidæ, Lucanidæ, Scarabæidæ, Buprestidæ, Elateridæ, Lampyridæ,* and *Ptinidæ.*

II.—**Heteromera.**—First and second pair of legs with five-jointed tarsi. The hind pair of legs with four-jointed tarsi. The most important families are the *Cantharidæ* and the *Tenebrionidæ.*

III.—**Tetramera.**—All three pairs of legs with four-jointed tarsi (*i.e.,* five-jointed, with penultimate joint rudimentary). The most important families are the *Bruchidæ, Curculionidæ, Scolytidæ* and *Cerambycidæ,* and *Chrysomelidæ.*

IV.—**Trimera.**—All three pairs of legs with three-jointed tarsi (*i.e.,* four-jointed, with penultimate joint rudimentary). The most important family is the *Coccinellidæ.*

Note.—The students should sketch the different forms of tarsi from specimens.

Cicindelidæ (Tiger-beetles).—These are brightly-coloured beetles, with long straight antennæ, large eyes, and powerful mandibles, which are set more or less vertically. They are fierce, active little creatures, which devour other insects. The rice sapper (*Leptocorisa*) is the only destructive Indian pest which seems to have been recorded as destroy-

Second Edition, revised. 8vo. Cloth. pp. xxx, 226.
Price 7s. 6d.

THE

MYSTIC FLOWERY LAND

Being a true Account of an Englishman's Travels and Adventures in China.

BY

CHARLES J. H. HALCOMBE,

LATE OF IMPERIAL CUSTOMS, CHINA.

WITH NOTES BY THE AUTHOR, AND NUMEROUS ILLUSTRATIONS.

"The book is one from which all may learn, but it is eminently suited to the 'general reader,' and its attractions are much increased by forty-one illustrations. The fact that it has achieved a second edition shows that it has already been received with favour, and we think that, in its present form, it should meet with a yet more hearty welcome."—*Sheffield Daily Telegraph.*

"The illustrations are all good, and the Chinese pictures reproduced in colours interesting. We have not seen any of them before."—*Westminster Review.*

LONDON: LUZAC & CO.
46, GREAT RUSSELL STREET (OPPOSITE THE BRITISH MUSEUM).

ed by tiger-beetles to any large extent in India, but other injurious species are also likely to be kept down by them. The larva is said to construct holes in the ground, where it lies in wait for passing insects. To aid it in climbing up the steep sides of its burrow, it has a pair of large hooks on the dorsal surface of the abdomen, which give it a curious hump-backed appearance. Some of the beetles mimic ants. The School Museum contains specimens of the beetle for the students to examine and sketch.

Carabidæ (Ground-beetles).—These insects are very numerous in India. They are much like Cicindelidæ in appearance, but can be readily distinguished by their smaller heads and horizontally set mandibles. Like the Cicindelidæ, they are predaceous in their habits and devour other insects. The larvæ of the migratory locust of North-West India (*Acridium peregrinum*, Oliv.) have been noticed as very largely destroyed by a species of Carabidæ belonging to the genus *Calosoma*, and other destructive insects also are likely to be kept down by them. The Museum contains a number of specimens for examination.

Dermestidæ.—These are small obscure insects with deflexed heads and weak legs. Most of the common Indian species are covered with pubescence. The larvæ are hairy grubs which attack dried animal and vegetable matter of all kinds. When full fed, the larvæ hide themselves away, sometimes burrowing into hard wood to obtain a safe retreat. They are a great nuisance in museums, and also do a good deal of damage to leather and other stored goods, including wheat. Skins that have been thoroughly treated with arsenical soap are not attacked by them. The Museum contains specimens of the species *Dermestes vulpinus* in various stages of development, which the students should examine and sketch.

Lucanidæ (Stag-beetles).—These are large insects, the males with enormously developed mandibles. They can be recognised by their peculiar elbowed antennæ with flattened terminal joints. The females are said to lay their eggs in the bark of trees near the roots. The larvæ are curved fleshy grubs with strong mandibles. Most of the species are believed to live in rotten wood, but the fact that ten per cent. of the oaks around Naini Tal have been reported as tunnelled by them shows that they also damage more or less sound wood. The life of the larva is said to last for from two to six years, but no very definite observations seem to have been made upon the subject in India. When full grown, the larva is said to construct a cocoon of chips in the burrow from which the beetle afterwards emerges. The beetles are said to feed on the sap which exudes from wounds on trees. The students should sketch a few representative specimens in the Museum, paying particular attention to the antennæ.

Scarabæidæ.—These beetles can be recognised by their peculiar antennæ, which have the terminal joints expanded into broad flat leaves, the basal joint not being enlarged as is the case with the Lucanidæ. The larvæ are soft curved grubs of very characteristic appearance. The family comprises an enormous assortment of insects including all the Rosechafers (*Cetonini*), Cockchafers (*Melolonthini*), Dung-beetles (*Coprini*) and Goliath-beetles (*Dynastini*). The cockchafers are mostly dull-coloured beetles of rounded outline. Their larvæ, which are the well-known *white grubs* of tea and coffee gardens, do an immense amount of damage by feeding upon the roots of plants of all kinds. In France they have been reported as one of the greatest pests in forest nurseries, and the same is likely to prove to be the case in India, though little has yet been recorded upon the subject. The beetles also do some damage by feeding on leaves and shoots. In Ceylon coffee estates, where cockchafer larvæ have been specially destructive, digging the insects out by hand is the only measure that seems to have proved at all generally practicable. With regard to the life history of cockchafers, the female beetle is believed to lay her eggs in the ground in the early part of the rainy season. From these eggs emerge the curved white larvæ which make their way through the soil and feed upon the roots of plants. In Europe they are said to require several seasons to arrive at their full growth, and the same will probably prove to be the case in India. When full grown they transform into pupæ underground, the beetles usually emerging at the close of the hot weather.

The students should also note the large Goliath-beetle (*Oryctes*), usually known as the "Rhinoceros-beetle" on account of the protuberance like the horn of a miniature rhinoceros which it bears on the upper part of the head. This beetle does a good deal of damage to palm trees by cutting large holes in the developing leaf-shoots. Its larvæ are to be found in heaps of rotting vegetable matter. The collection of these beetles by hand has been found worth doing in many cases, while the removal of heaps of rotting vegetable matter from the neighbourhood of palm trees is often recommended. In both Europe and Madagascar the larvæ of *Oryctes* beetles have been noticed as liable to be parasitised and killed by Hymenoptera of the genus *Scolia*, and the same is likely to prove to be the case with the Indian species. The School Museum contains numerous Scarabæidæ in various stages of development for the students to examine and sketch.

Buprestidæ.—These are mostly bright-coloured conspicuous insects of considerable size. They may be recognised by the straight serrated antennæ and large vertical head, which is sunk in the prothorax nearly up to the eyes. The larvæ are flattened footless wood-boring grubs, which can easily be recognised by the curious expanded prothorax. The female is believed to lay her eggs upon the bark, the larvæ tunnelling into the wood. Teak (*Tectona grandis*), sâl (*Shorea robusta*), khair (*Acacia catechu*), mango (*Mangifera indica*), and chir (*Pinus longifolia*) have been recorded as attacked by Buprestidæ larvæ in India, and other timber trees also are likely to suffer. But the insects appear to be far less numerous than the Longicorn larvæ which attack timber in a very similar manner. Damaged and unhealthy trees are the ones chiefly liable to attack, and the removal therefore of such timber from the forest is likely to reduce injury from these and similar boring insects. The School Museum contains a few Buprestidæ in various stages of development for the students to examine and sketch.

Elateridæ (Click-beetles or bow-and-arrow beetles).—These are generally small dull-coloured insects. They can be at once recognised by their characteristic shape. They have a peculiar spine on the ventral portion of the prothorax which is of assistance in leaping. The larvæ are known as wire-worms: they are little hard grubs not unlike pieces of slate pencil or bell wire cut into short lengths. The larvæ live in the ground, where they feed upon the roots of plants. As yet they have only been reported as injurious to potato plants in India, but considering how much damage they do to other plants in Europe and America, the pest is one that should be noticed. With regard to treatment, dressings of gas-lime, soot, nitrate of soda, salt, and rape-dust for the ground, have all been found useful in England, and are likely to be equally effective in India in cases where it is possible to employ them. The Museum contains specimens both of the beetle and the larva which should be examined by the students.

Lampyridæ. (Fire-flies).—These beetles can generally be recognised by their enormous eyes and soft cuticle. They comprise all the common species of Indian glow-worms and fire-flies, so should be noticed though they are of no practical importance. The larvæ are said to be carnivorous, but little is known about their habits in India. The phosphorescent apparatus is situated at the posterior of the abdomen, anatomically it is said to consist of a fat body thickly beset with tracheæ. The light therefore would seem to be produced by some process of slow oxidization.

Ptinidæ[1].—This family comprises a number of small obscure insects which do much damage by tunnelling into wood of all kinds. In the beetle the head is usually overhung in a peculiar manner by the prothorax, but this character is not always sufficient to determine them. The only family with which they are likely to be con-

[1] As defined by Packard, includes the *Bostrychidæ* of Gemminger and Harold.

founded is that of the Scolytidæ, which differ both in the structure of the tarsi[1] and in the possession of more produced heads. The larvæ of the Ptinidæ have legs, while those of the Scolytidæ are legless.

Ptinidæ larvæ tunnel into dead-wood of all kinds and, like almost all other boring insects, prefer for their attack wood in which the flow of sap is not abundant. Bamboos in particular are specially liable to be attacked and are often reduced to little more than powder. The bamboo insect is a species of *Dinoderus* which makes burrows about 1·5 millimetres in diameter. The eggs are believed to be laid by the mother beetle, soon after the bamboo is cut. The larvæ are little white grubs with powerful mandibles, which gnaw their way through and through the bamboos, the injury often remaining unnoticed until the larvæ have transformed into pupæ and the beetles begin to cut their way out. The best way to preserve bamboos from attack is to soak them in water for some time after being cut. This is said not only to prevent the deposition of eggs but to render the bamboos distasteful to the insect. For bamboos which have been thoroughly soaked in the first instance appear to be seldom attacked afterwards. Sponging bamboos over with kerosine oil is also an efficient prevention, when done early enough, but is only practicable in some cases.

The well-known cheroot weevil (*Lasioderma*) and book-worm (*Sitodrepa*), which are such a nuisance in India, also belong to the Ptinidæ, while stored grain, opium and numerous other substances are attacked by members of the same family. In Europe one of the best known representatives is the "death-watch" (*Anobium*), which is found in old furniture. The students should sketch one of the largest of the Ptinidæ beetles in the School Museum. They should also be shown bamboos attacked by the bamboo borer.

Cantharidæ (Blister beetles).—This family of heteromerous beetles comprises a number of brightly-coloured insects, the commonest of which (*Mylabris*) have the elytra banded with black and yellow stripes in a manner which renders them easily recognisable. The beetles have been reported as doing a little damage in India as defoliators, but are not likely to be of much practical importance. The larvæ do not seem to have been observed in India, but in the case of the common

[1] The Scolytidæ are said to have tetramerous tarsi while the Ptinidæ have pentamerous tarsi. The basal joint in the Ptinidæ, however, is often so minute as to be difficult to make out, while the rudimentary penultimate joint in the Scolytidæ is sometimes visible. By remembering, however, that in one case it is the basal joint which is rudimentary and in the other the penultimate joint, this difficulty disappears. The two groups should be carefully distinguished, as they have somewhat different habits. They have become a good deal confused with each other owing to the looseness with which the name Bostrychidæ has been employed by different writers.

black and yellow species they are likely to be white grubs which feed underground upon vegetable matter. In the case of one of the European species of *Meloë*, a genus which also occurs in India, the larvæ are parasitic upon bees and have a most complicated life history. Cantharidæ beetles are remarkable for their irritating secretions, which no doubt serve to protect them from being eaten by birds. In a dried state the beetles are used medicinally as irritants. The students should sketch a Mylabris beetle.

Tenebrionidæ.—This is a large family of heteromerous beetles. All the common species are blackish in colour and some of them are liable to be mistaken for Carabidæ. They can readily be distinguished from the Carabidæ by the structure of their tarsi. Most of the species are sluggish in their movements and many of them are said to have the power of emitting offensive secretions when disturbed. The larvæ appear to be mostly much like wire-worms (Elateridæ), only flatter. They live upon vegetable matter, including grain and flour. One species is said to tunnel into the stems of sandal-wood trees in Mysore, while another has been reported as injurious to agricultural plants, but little has been ascertained upon the subject. Tenebrionidæ beetles are excessively common in India, so should be noticed, though they are not known to be of any very special importance. The students should sketch one or two typical specimens from the School Museum.

Bruchidæ (Pea and gram weevils).—These tetramerous beetles are small, thick-set obscure-coloured creatures, with the front of the head produced into a short rostrum. The front wings do not cover the posterior of the abdomen. The only family with which they are likely to be confounded is that of the Curculionidæ, from which they can be at once distinguished by not having elbowed antennæ. Bruchidæ attack the seeds of leguminous plants of all kinds in India, and are often most destructive. The only tree which has been specially reported as suffering in India is the tamarind (*Tamarindus indica*), but the seeds of almost all forest representatives of the Leguminosæ are likely to be attacked to a greater or less extent. The common weevil which attacks gram may be taken as a typical representative of the family. The eggs are said to be laid in the developing pods. The larvæ are little soft footless grubs which bore into the seeds, which they often injure to the extent of preventing germination. The pupa is usually formed within the seed, a large hole being formed when the beetle emerges. It is said that seed affected by Bruchidæ larvæ can be disinfected without destroying its germinating power either by dipping it into boiling water, or by keeping it in some close-fitting receptacle with a little carbon bisulphide. Treatment of this kind, however, should only be attempted under careful suervision, and even then it would at first be of the nature of an experiment.

Creulionidæ (Weevils).—This family of tetramerous beetles comprises an enormous number of species which, one way and another, do a vast amount of damage in India. The beetles can be easily recognised, as they have the front of the head produced into a proboscis furnished with a pair of elbowed antennæ. The larvæ are white legless grubs, which tunnel into vegetable matter of all kinds. The palm weevil (*Rhynchophorus*) may be taken as a typical representative of the family. This insect destroys a large number of palm trees of all kinds. It is said to lay its eggs at the base of the leaf-stalk, on some spot where the stem has been injured, or in holes drilled by the rhinoceros-beetle (*Oryctes*). The larvæ tunnel their way through the heart of the trunk and often kill the tree outright. The pupa is formed in a cocoon of palm fibre, in the burrow. The beetles fly at night; in the day time they are often to be found concealed in the holes made by the rhinoceros-beetle. As with other wood-boring insects, unhealthy trees are likely to be more subject to attack than healthy ones.

The only remedies that seem to have been tried are the obvious ones of catching the beetles by hand and of burning trees that are badly infested by the larvæ, so as to check the increase of the insect. Trees that are not very badly affected should be spared, as they are said in many cases to recover.

Numerous timber trees in India have been reported as tunnelled by Curculionidæ larvæ, much in the way that palm trees are tunnelled by the species peculiar to them. We may notice:—the young mahogany trees destroyed by the larvæ of an undetermined weevil which tunnels beneath the bark, also chir (*Pinus longifolia*) and dhak (*Butea frondosa*) said to be tunnelled by species of *Astycus* and *Sipalus*. The asparagus-like shoots of the hill bamboo (*Dendrocalamus*) have also been reported as destroyed by a large species of the genus *Cyrtotrachelus*, and these are likely to be mere isolated instances of what goes on in connection with many other forest growths.

No less injurious are the weevils which attack seed and fruit in India. The seed-crop of sâl in the North-West Provinces has been reported as suffering to a very large extent [1] while comparatively recently, in Darjeeling, ninety per cent. of the seed of *Quercus pachyphylla* has been recorded as similarly lost.

[1] In 1867 Mr. R. Thompson wrote—"I have reported before in my letter No. 63, dated 1st September 1866, that entire seed-crops of the *Vatica robusta* were destroyed in 1863, and that partial destruction of them was observed since."

That these are merely typical examples of what is likely to go on generally is indicated by the fact that the injury by weevil to stored wheat and rice in India has been estimated at several millions of rupees annually. The fruit of the mango, again, in Bengal is very largely attacked by a weevil whose larva lives in the pulp and thus renders the fruit to a great extent uneatable. In the case both of seed and of fruit attacked by weevil, the hole made by the grub in cutting its way into the interior is so small that it is very difficult to detect injury until the beetles begin to emerge.

Wheat which is kept tightly packed in village pits, where the air is almost completely excluded, is found to escape attack. Seed also kept in a closed receptacle of any kind with a little bisulphide of carbon or naphthaline is also said to avoid injury. No satisfactory method of dealing with the mango-weevil has yet been discovered, but it may be noticed that in the United States, where plums suffer to some extent from an insect with very similar habits, spraying the trees with washes made of an arsenical poison known as London purple, in the early part of the season, has been found effective. This treatment would only be practicable under exceptional circumstances in India, and would always be somewhat dangerous, as London purple is excessively poisonous to all animals. The students should examine and sketch a few typical Curculionidæ from the Museum collection.

Scolytidæ (Bark borers).—This family of tetramerous beetles comprises numerous species which are destructive to forest trees in India. Conifers in particular are specially liable to attack. The beetles are small obscure creatures, which often have much superficial resemblance to the Ptinidæ. They can, however, be distinguished from Ptinidæ beetles both by the structure of the tarsi and also by the fact that their larvæ are without legs.

The female beetle usually bores a tunnel in the bark and lays her eggs at intervals along it. The larvæ on emerging from the eggs tunnel in all directions away from the original burrow, often travelling immediately between the bark and the wood, and leaving behind them a pattern of very definite character. When full grown they transform into pupæ, in the burrows, the beetles afterwards cutting their way out through the bark. Several generations are likely to be gone through in the course of the year.

Like other wood-boring insects the Scolytidæ are believed to prefer for their attack trees whose vitality has been impaired by injury or disease. Forest fires and the attack of leaf-feeding caterpillars are probably fruitful sources of the increase of bark-borers in Indian forests. In cases where Scolytidæ have become very numerous and where damaged trees are not available, they are said to attack healthy trees, and although the pioneer insects may perish through being suffocated in the sap, their burrows lower the vitality of the tree and fit it to harbour insects bred elsewhere.

In parts of Germany, in order to minimise damage by these pests, the plan is said to have been followed with some success of felling a certain number of trees in the spring, when the beetles are on the wing. The felled trees being in a withering condition, attract the insects and become the receptacles of vast numbers of eggs which can readily be destroyed by taking off the bark and burning it before the insects come to maturity.

Scolytidæ have been reported in India as attacking oak, pine and sâl, the damage in some cases being very considerable. It is one of this family also which drills the tiny holes so often noticed in beer barrels in India. In the case of the Conifer forests of the Sutlej valley, which have suffered to a large extent from the attack of a small Scolitid which is likely to be a variety of *Tomicus chalcographus*, Linn., Mr. Mein observed in 1887 that both *Pinus excelsa* and *Pinus Gerardiana* were attacked. The insect was chiefly active during the rains and disappeared almost completely in October. Larvæ were noticed in the tunnels about July. Both tender terminal shoots and large branches were attacked, the minute holes penetrating as far as the heart-wood. The only remedy attempted was that of lopping off infested branches and removing trees which had gone too far to be likely to recover. The students should examine the specimens illustrative of this group in the School Museum. In particular they should sketch the curious arrangement of the tunnels made in the wood.

Cerambycidæ (Longicornes).—This large group of tetramerous beetles contains a great number of insects which are extremely destructive in Indian forests. The beetles may be recognised by their enormously elongated antennæ and characteristic vertically set heads. They have a peculiar stridulating apparatus situated between the dorsal plates of the prothorax and the mesothorax. The larvæ are flattened legless grubs, with firm integument and powerful mandibles. The only wood-boring larvæ which they at all resemble are those of the Buprestidæ, and they can be at once distinguished from these by not having the prothorax specially flattened or disproportinately enlarged.

The larva in many cases passes a number of seasons in the wood before arriving at its full growth, a fact which is illustrated by the frequent emergence of the beetles from articles of furniture which have been made up for years. The mother beetle is said to lay her eggs in the bark, and after the tiny larva has tunnelled into the wood, there is often no visible sign by which to detect the presence of the insect until the beetle cuts its way out, though the larva can frequently be heard gnawing away at the wood inside. The pupa is to be found in the burrow near to the exterior.

The names of the forest trees that have been noticed as attacked by Cerambycidæ beetles in India are too numerous to record in this place. They include teak and sâl and many other common species. Indeed the probabilities are that hardly any timber is exempt. In the case of insects which were reported as damaging young teak trees in the Kulsi plantation in Assam, the injury could usually be detected by a swelling at the spot attacked. The measures adopted were to coppice young trees that were affected, old trees being left to themselves, as they often recovered.

A noticeable feature in connection with the species (*Xylotrechus quadrupes*, Chevr.) which at one time did an enormous amount of injury in Southern India by tunnelling into the stems of coffee-bushes, was the fact that the eggs required a good deal of sun-light to enable them to hatch. This was thought to explain the fact that estates which were well shaded escaped serious damage, while unshaded areas close by were largely destroyed.

The species *Plocederus obesus*, Gahan, which is common in various timber trees in Dehra, is remarkable on account of the calcarcus egg-like cocoon in which the pupal stage is passed.

Striking instances of the preference which the Cerambycidæ have for wood in a withering condition are afforded by the *Cælosterna* which has been reported as attacking coppice sâl saplings in Oudh; also by the species of *Sthenias* which attacks rose bushes and other shrubs in Southern India. Each of these insects first kills a branch by cutting a deep notch around it with its mandibles. It is then believed to lay its eggs above the place, thus securing for its larvæ the condition most favourable for their development.

The only general preventive measures that can be at all confidently recommended at present for keeping down Cerambycidæ beetles in a forest are the removal of dead and dying wood of all kinds, including fallen branches, and the barking of timber as soon as possible after it is felled. The School Museum contains very numerous specimens of Cerambycidæ beetles and their larvæ, also wood of all kinds tunnelled by them. The students should sketch one or two typical beetles and larvæ.

Chrysomelidæ.—This family of tetramerous beetles comprises a great number of little bright-coloured, thick-set, insects. They usually have the head partly sunk in the prothorax and differ totally in general appearance from the members of the other groups of tetramerous beetles which have been noticed in the preceding pages. • Both the beetles and their larvæ feed upon leaves, and several of the species do much damage in India by defoliating agricultural plants. In Europe and America they occasion some loss also by defoliating forest trees, and a species of the genus *Melasoma* has recently been reported as attacking willow trees in the North-West Himalayas. Generally speaking, however, the family has not been noticed as specially destructive in this connection in India. The larvæ are usually active little creatures, sometimes covered with spines on which they carry their cast-off skins as a kind of shelter. The pupæ may be formed either on the leaves or in the ground. Several generations are sometimes gone through in the rainy season. Spraying the insects with washes made of the arsenical poison known as London purple will always destroy them, but this treatment is only likely to be practicable in exceptional cases in India.

The School Museum contains specimens in various stages of development for examination by the students.

Coccinellidæ (Lady-bird beetles).—These trimerous beetles have often a good deal of superficial resemblance to Chrysomelidæ, but can be at once distinguished by the structure of their tarsi. They are small in size and usually more or less hemispherical in shape. Both in the larval and the beetle stage they feed upon Aphidæ and Coccidæ, and are most useful in keeping these destructive insects in check. Many species of Coccidæ have particular lady-birds which specially feed upon them. This fact has been taken advantage of with the best result by the United States Entomological Department, who have successfully colonised Coccinellidæ beetles in several parts of the world amongst destructive scale insects attacked by them.

In Dehra there is a common species by name *Epilachna viginti-octopunctata*, Fabr., which, contrary to the usual habits of the Coccinellidæ, feeds upon the leaves of plants. This case however must be looked upon as altogether exceptional.

The School Museum contains numerous species of the group for examination.

V.—Hymenoptera.

These insects pass through a well-marked pupal stage. They have the mouth parts formed for biting or licking. The wings, when present, are four in number and membranous, with comparatively few nervules.

On comparing the mouth parts of the bee with those of the cockroach we find that there is no very essential difference in the labium and the mandibles; the maxillæ are much produced to aid the tongue, which is composed of an elongation of the central portion of the labium, the labial palps being elongated and modified to assist. If time allows, each student should pick out, mount, and sketch the mouth parts of a bee, comparing them with his sketches of the corresponding organs in the cockroach.

The abdomen in the female usually ends with an ovipositor, which is often modified into a sting. In the latter case it is connected with a gland secreting a poison in which formic acid is probably the most prominent ingredient. The sting is a grooved spine enclosed in a sheath, from which it can be freely extended.

The Hymenoptera arose comparatively late in geological history, and comprise some of the most highly specialised insects. Some twenty-five thousand species are said to have already been described from different parts of the world, and vast numbers of minute forms, especially in India, are still unclassified.

The following are the more important groups :—

I.—TEREBRANTIA.—The female with ovipositor.

a. Phytophaga.—Abdomen sessile, the larvæ resemble caterpillars. Include the Tenthredinidæ.

b. Gallicola.—Abdomen stalked. The larvæ inhabit galls. Include the Cynipidæ and the Blastophaga.

c. Entomophaga.—Abdomen stalked. Larvæ parasitic on other insects.
Include the Ichneumonidæ and Chalcididæ.

II.—ACULEATA.—The female with ovipositor more or less modified to serve as a sting.
Include the Formicidæ, Chrysididæ, Apidæ, Vespidæ Fossoria, Scoliidæ, and other groups of less importance.

Tenthredinidæ (Saw-flies).—These Hymenoptera can easily be recognised by the fact that they have no constriction between the thorax and the abdomen. The larvæ resemble caterpillars in appearance, but can easily be distinguished from caterpillars by their indefinite number of prolegs and single pair of large ocelli.

Saw-fly larvæ feed upon the leaves of plants, and occasionally do a little damage in gardens as defoliators (*e.g.*, the species common upon rose bushes in the Dehra school compound). The eggs are laid in the bark of the twigs of the food plant, and result in a large wound, which may be injurious if the shoot attacked happens to be the leading one. Saw-fly larvæ can be destroyed by spraying with arsenical insecticides in cases where this treatment is worth adopting.

The rose bushes in the school compound usually afford a plentiful supply of saw-fly larvæ in various stages of development. The Museum also contains specimens of the imago and cocoon.

Cynipidæ (Gall wasps).—These are mostly minute fly-like Hymenoptera, which the students need only be able to recognise in connection with the galls that they make. The female deposits her eggs in the tissues of plants, where irritation is set up, resulting in a morbid vegetable growth, or gall, in which the larva lives. The larvæ themselves are thick-set fleshy grubs. In some of the European species the life history is complicated by asexual reproduction not unlike what goes on amongst the Aphidæ. The Cynipidæ are of no great importance to the forester, but should be noticed on account of their gall-making habits.

The School Museum contains specimens of Cynipid galls and of the insects which produce them.

Blastophaga (Fig insects).—These minute insects inhabit figs and are of considerable importance, as they are believed to be essential to enable reproduction to take place in many species of the genus *Ficus.* As observed by Cunningham in the case of *Ficus Roxburghii,* the females are winged, the males wingless. They are to be found sometimes in vast numbers in the interior of figs, and the students need only be able to recognise them in this connection.

The eggs are laid and the grubs develop in the ovaries of the aborted female flowers in the interior of the male fig capsule. Copulation takes place in the interior of the fig capsule, where the insects are bred. The male insects cut a passage out of the figs through the pollen-bearing flowers. After this the males die or are eaten up by ants, while the females escape by the passage. After flying about for a time the females are said to force their way into a young fig, which may be one containing either male or female flowers. Here the irritation which they set up by their movements, and possibly also the stimulus caused by the pollen which they are likely at least in some cases to bring with them from the male fig capsule where they were bred, cause the growth of the capsule and the development of seed. The females are only able to oviposit successfully in the male capsules, but neither the male nor the female capsules of *Ficus Roxburghii* can develop unless they are penetrated by the insect. Figs which are not so penetrated drop off and perish. The case of *Ficus Roxburghii* is believed to be typical of the relation between the Blastophaga and trees of the genus Ficus generally, in which case the part played by these insects in Indian forests must be very important. A plentiful supply of fresh figs infested with multitudes of Blastophaga can usually be picked up in the autumn from trees on the Rajpore Road in Dehra.

Ichneumonidæ.—These parasitic insects may be recognised by the long slender body and many-jointed antennæ. The female usually has a long protruding ovipositor. The egg is deposited by the parent insect in the tissues of a caterpillar. The larva is a little white legless grub which feeds upon the fatty tissues of its host, while the latter eventually dies of exhaustion, though it may have strength before this happens to transform into a chrysalis. When full fed, the Ichneumonid larva spins itself up into a cocoon, which is often attached to the body of the dead caterpillar.

Very many caterpillars in India, including numerous wild silkworms, are liable to be parasitised by Ichneumonidæ, and the group must be looked upon as of some importance to the forester, as tending to keep down defoliating caterpillars in the forest. It should be noticed, however, that the Ichneumonidæ are very much less numerous in India, and consequently less effective than such insects as the Tachinæ, which attack caterpillars in a very similar manner.

The School Museum contains numerous specimens for examination.

Chalcididæ.—In size and build these small insects are superficially not unlike house-flies, which belong to the totally different order of Diptera. As their name implies, the Chalcididæ are generally more or less metallic in coloration. The wings are usually somewhat deficient in veins. The antennæ have from six to fourteen joints. In some of the species the posterior femora are very much expanded. The Chalcididæ are mostly parasitic on caterpillars, and in habits they resemble the Ichneumonidæ. They have been reared from more than one forest-defoliating caterpillar in India, and no doubt have some effect in keeping insects of this kind in check. So far as has been observed, however, they are very much less effective than such insects as the Tachinæ.

The Museum contains a few specimens.

Formicidæ (Ants).—Ants are excessively common in India. They may be recognised by the basal segment of the abdomen, which is constricted off so as to form as it were a knot in the stalk by which the posterior portion of the body is attached to the thorax. Care should be taken not to confuse the ants with the totally distinct insects which are known as white-ants (*Termitidæ*), and which belong to the Orthoptera (*vide* pages 12 and 16).

Ants live together in communities, which contain winged males and females and wingless neuters. The neuters are supposed (like worker bees) to be aborted females. They are usually of two kinds, which are distinguished as the worker major and worker minor. They possess poison glands and are said to be either armed with stings or to inject the poison fluid into wounds made with their mandibles.

Ants form nests, which consist of passages and chambers dug out in earth or rotten wood, or built in leaves fastened together for the purpose. The larvæ are little legless grubs which are carefully looked after by the workers and are said to be fed with food elaborated in the stomachs of the latter. The food usually consists of animal matter, sugar, honey, or seeds. The seeds are believed in some cases to be converted to a great extent into sugary matter by fermentation before being eaten. The pupæ are enclosed in little silken cocoons (usually known as ants' eggs). Both larvæ and pupæ are often carried about by the workers from one nest to another, so as to secure for them the proper amount of warmth and moisture. The winged males and females fly out from the nest, generally after rain. When fertilised, the female drops her wings and returns to the old nest or starts a fresh one for herself and begins egg-laying. The female may live through several seasons, but the male dies off rapidly.

The most important sub-families of Indian ants are:—(*a*) the *Formicides*, with but one knot in the peduncle, and abdomen not furnished with a sting; and (*b*) the *Myrmicides*,[1] with two well-marked knots in the peduncle, the abdomen being usually furnished with a sting. Amongst the *Formicides* the students should notice—(1) the large fierce bamboo ant (*Œcophylla smaragdina*, Fabr.), with yellow workers and green males and females. It is plentiful in Dehra, so specimens can usually be found in large numbers for examination. It builds its nest in trees by joining leaves together, and feeds largely on caterpillars. The worker is mimicked in the most wonderful manner by a little jumping spider. The spider is said to feed on bigger and more active insects than the ants, and to utilise its resemblance to the ants to approach within springing distance of its prey. (2) The large black ant (*Camponotus*) also common in the school compound. This insect forms its nest in the earth and frequently attacks other ants. It is to be found in attendance upon Aphidæ, Lycænidæ caterpillars, and upon the curious little Homopterous insect (*Centrotus*), which are all believed to yield it certain sweet secretions. Its relations with some of the Lycænidæ caterpillars are most complicated.

Amongst the *Myrmicides* the students should notice—(1) the large fierce insectivorous ant (*Sima rufo-nigra*, Jerdon), with red thorax and black head and abdomen. It is to be seen upon almost every tree trunk in the school compound, but must be handled with care as it possesses a most virulent sting. It forms its nest in dead-wood. It is mimicked in the most wonderful manner by a little red and black spider, specimens of which are preserved in the Museum. (2) *Solenopsis gemminatus*, Fabr. This is the common red ant

[1] The remaining sub-families are:—(*a*) *Ponerides* with one well-marked knot in the peduncle followed by a second imperfectly marked knot, the abdomen with sting; and (*b*) the *Dorylides*, an altogether anomalous group, which may be easily recognised by the large yellow-winged males and curious little flattened yellow workers, the females being unknown.

of the plains of India. It forms its nests, which are very populous, in the ground, usually under stones or bricks. It constructs partially covered ways across roads, and is often to be seen carrying off dead insects. It has been said to attack potatoes in the ground, and is likely to do some damage in nurseries. (3) *Holcomyrmex scabriceps*, Mayr. This is a largish ant, with reddish head and dark abdomen. It forms its nests in the ground, and stores up grass and other seeds. Quite a large heap of chaff is often to be found outside the nest. The granary inside is sometimes as big as a walnut. According to Walsh's observations, the object in storing is likely to be, not so much to provide food for future use, for the ant is nearly torpid all through the hot season, as to allow fermentation to set in, thus converting the starch into sugar. The seed does not germinate in the granary, though it germinates freely outside. To prevent germination the ants not only take great pains to keep the seed dry, but are believed to bite off the radicle. (4) A species of *Cremastogaster*—a genus which may always be recognised by the curious pointed abdomen—has been reported as a great nuisance in Ceylon, as it builds large nests, sometimes more than two feet in diameter, in coffee and cinchona bushes, and bites so sharply that it is difficult for coolies to work anywhere in the immediate neighbourhood. Allied species of the same genus are common in the school compound.

The studen ts should sketch the arrangement of the knots on the peduncle which charac terises the various sub-families. For this purpose specimens of *Camponotus* and *Sima* may conveniently be taken to show the difference between the Formicides and Myrmicides.

Chrysididæ (Cuckoo-wasps)).—These are little metallic-green creatures with thick armour. They may be recognised by the basal rings of the abdomen, which are enlarged so as to enable the posterior segments to be retracted within them. The abdomen beneath is concave and the insect can roll itself into a ball when disturbed, a habit which is often of use as a protection against the Hymenoptera, in whose nests it is parasitic. The larvæ are legless grubs, which feed upon the young of other Hymenoptera.

The Museum contains specimens for examination by the students.

Apidæ (Bees).—Bees can be recognised by their hairy bodies and expanded posterior tibiæ, which are modified to serve as pollen receptacles, the basal tarsal joint being enlarged to assist. The students should sketch the hind leg of a bee to remind them of its peculiar structure. Bees feed upon honey and pollen. Their chief importance in the forest consists in the fact that they serve very largely to convey pollen from flower to flower. Some of the social species are also of some slight importance as producers of honey and wax. The family may conveniently be divided into (*a*) social bees, and (*b*) solitary bees.

Amongst social bees the community consists of a female or queen, a number of males or drones, and a host of aborted females, which are known as workers and which do all the labour of the hive. The honey is collected by the workers from flowers. It is said to be partially acted upon in the stomach of the worker and thence regurgitated into the cells. The young are fed by the workers upon a mixture of honey and pollen. The wax is excreted in thin plates from glands which are situated on the ventral surface of the body between the segments. The comb is built up by the workers in the shape of more or less regular hexagonal chambers. The cells are used partly for storing honey and pollen and partly for rearing larvæ. Only one adult queen is to be found at one time in the nest. When a queen emerges she takes flight and is fertilised by a drone. She then returns to the nest, where she lays her eggs. In the meantime the old queen leaves the nest with a number of old workers to found a new colony, thus constituting what is known as a swarm.

Amongst the social bees the students should notice *Apis dorsata*, *Apis mellifica*, *Apis indica*, *Apis florea*, and *Trigona sp.* *Apis dorsata* is the large jungle honey-bee, which builds a single comb in the open attached to tree branches or rocks. It can be distinguished by its size and the elongation of its body. Both the honey and wax produced by it are collected for sale, and in some cases yield a small return to the Forest Department, though the honey is somewhat poor. The bee stings badly, and from its habit of attacking *en masse* is occasionally dangerous to man. Specimens are preserved in the School Museum. *Apis mellifica* is the common European hive bee. A variety of it is cultivated in the hills of the Punjab and Kashmir, where it yields a large amount of excellent honey. It is somewhat smaller than *Apis dorsata* and can readily be distinguished by its shorter and thicker build. *Apis indica* is the common bee of the plains. It is almost exactly like *Apis mellifica* in shape and coloration, but can be distinguished by its smaller size. It builds its combs in cracks and crannies, and is said to be occasionally kept in a state of partial domestication in earthen pots or in holes made for the purpose in house walls, but it yields very little honey. Specimens of it are preserved in the School Museum. *Apis florea* is much smaller even than the *Apis indica*. It may be recognised by the reddish colour of the basal segments of the abdomen. It builds a small single comb in the open upon the branches of trees, and stores very little honey. It is mimicked in a wonderful manner by a solitary bee of the genus *Sphecodes*, which is parasitic in its cells. Besides the various species above enumerated of the genus *Apis*, there are in India several species of the genus *Trigona*, a group of small stingless bees which build in old walls. They form their nest chiefly of chewed resin. The honey is said to be stored in large cells in the bottom of the nest. The species *Trigona vidua*, St. Farg., has been observed by Major Bingham to be mimicked most accurately by an Asilid fly (*Diptera*), which, though of almost exactly the same size as the bee, is able to seize and devour the latter. The students should be able to distinguish between the various honey-bees in the school collection.

In India honey-bees have many enemies. The caterpillar of the little wax-moth (*Galleria*) tunnels through and through the combs, making silken galleries in all directions and often causing the nest to be abandoned. Innumerable predaceous animals, including wasps, Asilid flies, toads, and numerous birds, devour the bees, while diseases such as *foul brood* and parasites like *Stylops* and *Braulina* are also likely to be present, though they do not appear as yet to have attracted much attention in India.

Amongst solitary bees the commonest forms in India are the mason-bees (*e.g.*, *Megachile*) and the carpenter-bees (*e.g.*, *Xylocopa*). Mason-bees are to be seen at work in dry weather in verandahs almost everywhere. They build cells of leaf or mud and store them with honey and pollen. The female deposits an egg in each cell and carefully closes it up, leaving the larva to develop by itself. Carpenter-bees bore into old dead timber, where they form cells much like those made by the mason-bees. In this way they may occasionally do a little damage. Mr. Thompson notices that wood which has been painted over with tar or resin is never attacked, and that an application of tar to a post which is already affected will make the bees leave it. The larvæ of solitary bees are exceedingly liable to the attack of parasitic insects of many kinds. The subject is one of very considerable interest, but is not of sufficient importance to be worth going into at length in this place. The Museum contains numerous solitary bees for examination. The insects are also to be seen at work.

Vespidæ (Wasps).—These are smooth-bodied insects with elbowed antennæ. The front wing can be folded upon itself longitudinally. (The students should verify these points for themselves.) As in the case of bees, the Vespidæ may conveniently be divided into (*a*) social species, and (*b*) solitary species.

Social wasps live together in communities, which consist of males, females and workers. They build nests of a paper-like substance formed of chewed wood. The young hang head downwards in the cells and are fed as they grow by the wasps, chiefly on chewed insects. The fullgrown insect itself feeds chiefly upon sugary matter or fruit. The commonest Indian wasps are *Polistes hebræus*, which is the little yellow wasp with long legs common in verandahs, *Vespa affinis* and *Vespa velutina*, which build enormous nests in the jungle, and *Vespa magnifica*, a large insect which builds in hollow trees and has a most virulent sting.

Solitary wasps build cells much like those made by solitary bees; they store these cells with insects or spiders which they paralyse, but do not kill, by stinging them. When the cell is stocked, and the egg laid, the parent wasp closes it up and leaves its larva to develop alone. The parents are said to die off before the young emerge from the cells. The students can usually see solitary wasps at work building and stocking their nests in the school compound.

Fossoria (Sand-wasps).—This group comprises several families. The insects are smooth-bodied, often long-legged creatures, much like wasps in appearance. They can be distinguished from wasps by their antennæ, which are often curled but not elbowed. In habits they are very much like solitary wasps. They may frequently be seen dragging along paralysed grass-hoppers and other insects as big or bigger than themselves, in order to store them in cells for their young to feed upon. The Museum contains numerous specimens.

Scoliidæ.—These insects are hairy creatures, frequently of considerable size. They often have much superficial resemblance to bees, but can be readily distinguished from these insects by their posterior tibiæ and tarsi, which are not modified for pollen-carrying. Little has been observed in connection with their habits in India, but both in Europe and Madagascar representatives of the family have been observed to lay their eggs in the larvæ of the rhinoceros-beetle (*Oryctes*). It is likely therefore that the species of Scoliidæ which occur in India will be found to have similar habits. In this case they must be to some extent beneficial, as every beetle larva that is parasitised perishes. The School Musuem contains specimens for examination.

VI.—Diptera.

The Diptera comprise the two-winged flies. They may be defined as insects which pass through a well-marked pupal stage, have their mouth parts modified for piercing and sucking, the hind wings rudimentary, and the front wings, when present, membranous. There are a great number of species in India, many of the forms being vastly numerous in individuals. One of the easiest features for recognising a dipterous insect is the peculiar little pair of rudimentary hind wings which are present in the great majority of species and which are said to serve as organs of sense.

The mouth parts consist of the labium, which usually forms a sheath for the style-like mandibles, maxillæ, and hypopharynx. These style-like piercing organs are much developed in insects like the mosquito and the flea, which live by blood-sucking, but are rudimentary in the house-fly, which is therefore only able to suck up fluid with its flexible tubular labium.

Diptera are usually classified into the groups *Pupipara, Brachycera, Nemocera* and *Aphaniptera.*

(*a*) Pupipara.—Antennæ small or rudimentary. Flattened spider-like insects, many of them wingless. The young are said not to be brought forth until they have reached the pupal stage.

E.g., Hippoboscidæ.

(*b*) Brachycera.—Antennæ short, usually composed of not more than about four joints, mostly stout, thick-set winged insects.

E.g., Asilidæ, Tabanidæ, Œstridæ, and Muscidæ.

(*c*) Nemocera.—Antennæ long and many jointed. Slender, winged insects.

E.g., Culicidæ and Cecidomyidæ.

(*d*) Aphaniptera.—Degraded, wingless, laterally compressed insects, with short antennæ and distinctly separated thoracic rings.

E.g., Pulicidæ.

Hippoboscidæ.—About the commonest representative of this family in India is the horse-louse. This is the well-known flattened insect which looks very much like a winged spider and is exceedingly difficult to kill. It is often most annoying to horses, upon which it settles to suck blood. Allied species are parasitic upon birds and other animals.

Asilidæ (Robber-flies).—These flies can be recognised by their elongated hairy bodies and powerful claws. In the winged stage they are rapacious in their habits and devour great numbers of other insects. Little is known of the larval stages of the species which inhabit India, but they are likely to be white legless grubs which feed upon vegetable matter. The School Musuem contains specimens for examination by the students.

Tabanidæ (Gad-flies).—These well-known insects can be recognised by the characteristic shape of the head, and the structure of the terminal joint of the antennæ, which is annulate and without an appended bristle. In the female the mouth parts, especially the mandibles, are much developed, and the insect is able to inflict a formidable bite. The female sucks blood, while the male takes little or no nourishment after it reaches the winged stage of its existence. Little is known about the larvæ of the Indian species, but they are likely to be very variable in their habits. The School Musuem contains specimens for examination by the students.

Œstridæ.—These flies are mostly bright-coloured hairy creatures, with rudimentary mouth parts and minute antennæ. The larvæ of the species *Pharyngobalus cameli* are said to make their way into the pharynx of the camel, where they feed upon the mucous membrane, and produce great irritation of the nostrils and pharynx, causing an offensive discharge tinged with blood. When full fed the larva makes its escape through the nostril and falls to the ground, where it passes through the pupal stage. The species *Gastrophilus equi*, a variety of which is common in Dehra and probably throughout India, is said to lay its eggs, four or five hundred in number, generally on or about the horse's front legs. The horse takes the eggs into its mouth in biting the irritated spots where they have been laid. The larvæ thus pass into the stomach, where they attach themselves to the mucous membrane. Here they are said to remain for nine or ten months, feeding on the matter produced by the irritation they set up. When full grown they loosen their hold and are carried through the intestines and ejected with the excrement. They then burrow into the ground, where they transform into pupæ, the imago emerging in thirty or forty days. When only a few grubs are present in a horse's stomach they are said to cause but little injury to the animal, but when they exist in large numbers they may produce very considerable inflammation and even death. The Musuem contains a specimen for examination.

Other species, related to the European bot-fly (*Hypoderma bovis*, De Geer), lay their eggs in the backs of cattle, deer, goats and other animals. The larva penetrates beneath the skin and forms a sore, which does not heal until the insect is full grown and drops out to transform into a pupa in the ground. It is said that hides received from the North-West Provinces of India for shipment to Europe are often damaged to a large extent in this way. In the case of domestic animals, filling up the sore with a plug of almost any kind of greasy ointment, which should, when possible, be of an antiseptic nature, is said to suffocate the grub and cause it to release its hold, thus enabling the sore to heal.

Muscidæ.—This family comprises a vast number of species, with very various habits. The common house-fly may be taken as a typical representative. The antennæ are three-jointed, the terminal joint flattened and furnished at the side with a large bristle. The proboscis ends in a fleshy lobe. Amongst the Muscidæ the most important group of species are the Tachinæ, which have very great resemblance to the house-fly group, but differ in having the bristle at the end of the antenna smooth instead of plumose. The eggs of the Tachinæ are laid upon caterpillars and other soft-bodied insects. The larva (maggot) burrows into the caterpillar's body, but avoids the vital organs, so that the caterpillar usually lives on until the maggot is full fed and cuts its way out. On emerging from the caterpillar the maggot buries itself in the ground, where it transforms into a little brown bean-shaped pupa, from which the fly afterwards issues. Tachinid flies are of very great importance on account of the wonderful effect which they have in keeping down defoliating caterpillars of all kinds. They are excessively numerous in India, and every single caterpillar which is parasitised by them perishes before it reaches the stage in which it can reproduce itself. They are also a serious nuisance to silk-rearers, and special precautions have to be taken to prevent their destroying silk-worms wholesale.

In the case of the house-fly (*Musca domestica*) the eggs are laid in horse-dung and other decaying matter, and as a generation of the insect, under favourable circumstances, only takes about a fortnight, multiplication is excessively rapid. House-flies are believed to be frequent carriers of infectious diseases. We may also notice the mango-fly (*Dacus ferrugineus*, Fabr.), the peach-fly (*Rivellia persicæ*, Bigot), and the melon-fly (*Carpomyia parctalina*, Bigot) which lay their eggs in ripening fruit, and occasionally do a good deal of damage in India. Another important species is the locust-fly (*Anthomyia peshawarensis*, Bigot) which is parasitic upon the eggs of the migratory locust (*Acridium peregrinum*, Oliv.) of North-West India, and is believed to have a very considerable effect in keeping this destructive insect in check. The School Museum contains various specimens illustrative of the Muscidæ for examination by the students.

Culicidæ.—Mosquitoes lay their eggs on the surface of stagnant water. The larvæ are active little aquatic creatures, with tracheal breathing organs at the posterior of the body. They are usually to be found in great numbers in dirty water that has been allowed to stand for a few days. They are said to form one of the chief food supplies of young tank fish in India, and may possibly have some slight effect in preventing stagnant water from getting foul, as they feed upon the organic particles it contains. The pupæ live in water and, unlike this stage in typical Diptera, have considerable powers of movement. They have respiratory tubes connected with their thoracic stigmata, through which they breathe air at the surface of the water. The female sucks the blood of mammals and birds and is said by some writers to be the host of the injurious Nematode blood parasite *Filaria sanguinis hominis.* The male has mouth parts unsuited for blood sucking, and is said to take no food during the imago stage.

Under favourable circumstances mosquitoes can develop in about three weeks in India from the egg to the winged form, and as each female is capable of laying some hundreds of eggs, multiplication goes on excessively rapidly. Stocking tanks with fish is likely to tend to reduce the number of mosquitoes bred in the water. In the case of stagnant pools close to houses, which are too small to be worth stocking with fish, a little kerosine-oil poured upon the surface, over which it forms a delicate film, is said to be surprisingly effectual in preventing the breeding of mosquitoes.

The students should examine some mosquito larvæ and make out the breathing organs. They should also notice the structure of the antennæ in the winged form, both male and female.

Cecidomyidæ.—This family comprises a large number of small slender-bodied species, which lay their eggs in the tissues of plants and often produce galls, much in the manner that galls are produced by insects of the Hymenopterous family *Cynipidæ*. The destructive Hessian fly (*Cecidomyia destructor*, Say.) whose larvæ are parasitic upon wheat-stalks both in America and Europe, belongs to this family, while in India there is an allied species (*Cecidomyia oryzæ*) which attacks rice in a similar manner. None of the Cecidomyidæ have yet been recorded as destructive in Indian forests, so the students will not be expected to know much about them.

Pulicidæ.—Fleas lay their eggs in dust and dirt in houses. The larvæ are tiny little hairy grubs, which are to be found in floors of badly-kept rooms, where the pupal stage is also spent. The insect is said to pass the winter in the larval stage, but in warm weather less than a month is required for development from the egg to the imago. Burning the matting, cleaning up the floors, with a free use of Persian insect powder should be resorted to in houses infested by this pest.

VII.—Lepidoptera.

This order comprises the butterflies and moths, which are exceedingly numerous in India. They may be defined as insects which pass through a well-marked pupal stage, and have scale-covered wings and suctorial proboscis in the form of a coiled tube. The proboscis is the characteristic feature of the mouth of a typical Lepidopterous insect. It is composed of the blades of the maxillæ, which are much elongated and appressed together, so as to inclose a canal through which fluid can be sucked up by the action of the peculiar stomach with which the adult insect is furnished. On either side of the proboscis is usually a large hairy labial palp, the rest of the mouth parts being rudimentary.

In the winged stage, Lepidoptera feed largely upon the honey which they collect from flowers, they thus become important agents for distributing pollen from flower to flower.

The larvæ are worm-like creatures, known as caterpillars, which feed upon leaves. For an account of the structure of a typical caterpillar, see the chapter upon the dissection of the mulberry silk-worm. The only insects with which caterpillars are liable to be confounded are the larvæ of saw-flies (*Tenthredinidæ*). They can be distinguished from these insects by the possession of numerous ocelli situated on the lower part of the head and by the nature of the fleshy prolegs, which are not more than ten in number and are never to be found upon either of the two basal segments of the abdomen.

The caterpillar moults its skin several times in the course of its growth, and on the last moult emerges as a pupa (chrysalis), from which the butterfly or moth, as the case may be, afterwards issues. When full fed the caterpillar usually spins itself up into a silken cocoon, or hides under ground or in some sheltered corner, where it remains during the helpless pupal stage. Both butterflies and moths are exceedingly numerous, and it will only be necessary for the students to notice a few of the more important groups.

The following may be taken as a convenient classification :—

I. Rhopalocera (Butterflies).—Antennæ knobbed. Front wing not attached to hind wing—

(*a*) **Suspensi.**—Chrysalis naked and suspended.
E.g., Nymphalidæ.

(*b*) **Succincti.**—Chrysalis girt with a silken cord.
E.g., Papilionidæ and Lycænidæ.

(*c*) **Involuti.**—Chrysalis rolled in a leaf or other covering.
E.g., Hesperidæ.

II. Heterocera (Moths).—Antennæ not knobbed. Forewing usually attached to hind wing.[1]

(*a*) **Bombyces.**—Thick-set moths of considerable size, with pectinated antennæ.
E.g., Lasiocampidæ group, Psychidæ group, Hepialidæ group, and Bombycidæ group.

(*b*) **Noctues.**—Thick-set moths of considerable siez, with filiform antennæ.

(*c*) **Geometres.**—Slender moths often of considerable size with more or less pectinated antennæ.

(*d*) **Microlepidoptera.**—Minute moths, very various both in structure and in habit.

Nymphalidæ.—This family of butterflies may be recognised by their feeble forelegs, which are very imperfectly developed in both sexes and are not used in walking. The students should be able to recognise the following forms in the School Museum :—(1) The common black and white *Danaids,* which are not eaten by birds on account of their offensive flavour. They are mimicked in the most wonderful manner

[1] The School Museum possesses a specimen prepared by Mr. Clifford to enable the students to see the method of attachment of the wings to each other. Sketches should also be made by the students of the various forms of antennæ.

by butterflies of the genus *Papilio* (also to be seen in the collection), which belong to the totally distinct group of Succincti. The mimicking species have no offensive flavour and would be readily eaten by birds if it were not for their wonderful resemblance to the protected *Danaids.* (2) The leaf-butterfly (*Kallima*), which looks almost precisely like a leaf when at rest upon a bush, but which exhibits bright colours when upon the wing. (3) *Hypolimnas bolina,* Linn., the male of which exhibits bright metallic colour when flying towards its mate, but appears dusky and inconspicuous when pursued by birds.

Papilionidæ.—The butterflies of this family have the fore legs in both sexes fully developed and capable of being used for walking. Many of their caterpillars are provided with curious tentacle-like processes, which can be protruded with great rapidity from the front of the body. These appendages are said to be scent organs and are believed to be utilised in driving off Ichneumonidæ, which are one of the chief enemies of the caterpillars. The students should be able to recognise in the School collection—(1) the species *Papilio erithonius,* Cramer, whose caterpllars are very destructive to the foliage of young orange and lemon trees in India; (2) *Papilio polytes,* which is interesting on account of its three distinct forms of female; (3) the white butterflies of the group Pierinæ, whose caterpillars defoliate cabbages and other garden plants.

Lycænidæ (Blues and coppers).—The butterflies of this family are small slender insects, with somewhat weak forelegs. In the male the tarsi of the forelegs are long and without joints. In the female they are jointed, as in the two posterior pairs of legs. The students should notice the species *Catochrysops pandava* in the School collection, whose caterpillars, like others of the same family, are attended and protected by ants; also *Virachola isocrates,* Fabr., whose caterpillars tunnel into pomegranates and are said sometimes to fasten the fruit with silken strands on to its stalk so as to prevent its falling to the ground before the butterfly emerges.

Hesperidæ (Skippers).—The butterflies of this family may be recognised by the curious little hook at the end of the clubbed antenna. The caterpillars of the species (*Suastus gremius,* Fabr.) occasionally injure the foliage of palm trees, but are of no great importance.

Lasiocampidæ group.—This group comprises several families.[1] The moths are mostly fluffy inconspicuous creatures, which lay their eggs upon the leaves or bark of plants. The caterpillars are covered with hair and in many cases do a great amount of damage in forests by defoliating the trees. The hairs are often sharp and brittle, with considerable urticating properties. When full fed the caterpillars spin

[1] *Postscript.*—The Lasiocampidæ, Lymantriidæ, Eupterotridæ, etc., as given by Hampson, Vol. I.

themselves up into cocoons, which are usually formed of their own hair fastened together with a little silk. The cocoons are to be found either attached to the leaves or spun up in the litter beneath the trees. Very extensive areas of forest in India are sometimes defoliated by these insects, with the result, not only of a considerable loss of growth, but also of injury to the vitality of the trees, which renders them liable to the attack of Scolytidæ and other boring insects.

Sâl forests are specially liable to attack. In 1878 the sâl trees throughout two hundred square miles of forest in the Dooars are said to have been completely defoliated by caterpillars of the genus *Dasychira*, which may be recognised by the four thick tufts of hair set in the middle of the back of the anterior segments of the abdomen. This instance may be looked upon as a typical one. The insect mysteriously disappeared almost as suddenly as it came, owing, it is believed, to the attack of the Tachinid parasites (*Diptera*) to which it is known to be very subject. In India Tachinæ are believed to be about the most important agents in keeping down the numbers of defoliating caterpillars of this kind, but Chalcididæ and probably also Ichneumonidæ (*Hymenoptera*) do useful service. In the case of the *Nun* (*Liparis monacha*) caterpillar, which has occasioned an enormous amount of damage to spruce in the forests of Central Europe, besides attacking beech and Scotch fir, a disease apparently identical with the *flacherie* of silk-worms, is said to have ultimately proved a most effectual natural remedy, though, in the last attack, it appeared too late to prevent injury on a vast scale, as the spruce fir over very large areas died from the destruction of the foliage. The fact is an interesting one, as *flacherie* is a disease which attacks both mulberry and Eri silkworms in India.

The most successful artificial measure adopted in Central Europe against the *Nun* caterpillars seems to have been that of localising the attack by forming protecting belts around infested areas. In the *Indian Forester* of July 1891, Sir D. Brandis gives an interesting account both of the attack and also of the remedies adopted in different localities. Speaking of what was done in 1890 in the "Dürrenbucher Forst" on the banks of the Danube, where the forest was a mixed one, the undergrowth largely consisting of spruce, he writes :—

"On the protection belts, which were made 100—130 feet wide, three operation were undertaken. *First*, on all old stems rings, twelve inches broad, of tar and glue were applied at a convenient height, about four feet from the ground. This mixture of tar and glue has for some years past been specially prepared, and has, as previously mentioned been employed on a large scale against the ravage of the Scotch pine moth, the well-known *Gastropacha pini*.[1] To make the glue stick on the stems, the rough outer bark is scraped off with a sharp knife, which gives a smooth surface to the inner reddish bark Hence this operation is commonly designated as *rothen*, to redden the stem. These rings or belts of glue, if the mixture is good, remain

[1] One of the same group of moths with hair-covered defoliating caterpillars.—E. C. C.

sticky during two or three months, and they effectually prevent the caterpillar from ascending the tree. *Secondly,* all underwood was cut and burnt as far as it could not be utilised. Poles fit for sale were barked and the bark was burnt with the rest. For this purpose special fire places were prepared and surrounded by a circular ditch with steep sides, to prevent the spread of the fire and the escape of the caterpillars, which, as soon as the underwood was heaped up inside the ditch, rapidly collected in the ditch and were destroyed. The moths commenced to swarm in this district, while huge masses of caterpillars were still at work feeding. Hence the fires lit to burn the underwood were also used to attract and destroy the moths; for this purpose the fires were lighted at nightfall and kept up until midnight. Between nine o'clock and midnight the largest swarms of the moths appeared. A portion were actually burnt, the others were stupefied by the heat, fell to the ground, and were easily killed. Pine torches were also used with large screens behind them, smeared over with glue, on which the moths were caught. *Thirdly,* extraordinary efforts were made to destroy as many caterpillars and moths as possible. Gangs of women with blunt brooms and long handles were busy in the protection belt and elsewhere in the forest, in killing the caterpillars, which were set in motion by the cutting of the underwood and other operations undertaken. For the destruction of the moths 300 labourers and 80 boys under their masters were employed. This work could, as a rule, only be done early in the morning from four to half-past eight o'clock. Later in the day the moths fly about and cannot be caught. Only on cool rainy days do the moths remain quiet during the day and can be destroyed.

"To return to the protection belts, the great object aimed at was to stop the spread of the caterpillars by making it impossible for them to ascend large trees, and by destroying the food which the underwood had furnished them. Here and there the success was not complete, and in such cases an additional protection belt was formed outside the one first established. Upon the whole, the result was most satisfactory, and the destruction was confined to less than a sixth part of the entire forest. In some cases ditches with steep sides were used in addition to the protecting belts, and the caterpillars collected in them in large masses."

It should also be noticed that all leaf-feeding caterpillars can be destroyed by spraying the trees with arsenical insecticides. This form of treatment is being largely adopted for fruit trees in America and Europe. It is only likely to be of use in the case of specially valuable fruit trees in India.

The School Museum contains numerous specimens in various stages of development for the students to examine.

Psychidæ group (Bag-worms).—These insects defoliate plants of many kinds in India and do some damage in forests. The caterpillars may be recognised by the curious little cases composed of stick or leaf which they carry about upon their backs. The male is an obscure little moth, which is so active in its habits and lives such a short time that it is seldom noticed. The female is a degraded, wingless creature, which never quits the case that is formed by the caterpillar. When the caterpillar is full fed it attaches the case by a silken strand to a leaf or twig. It then fastens up the open end with silk and transforms into a chrysalis. The female is fertilised by the male and lays all her eggs in the case. In collecting bag-worms by hand, therefore as is often done upon tea gardens, where they do a good deal of damage, care should be taken to collect all the apparently dead cases, as these are always liable to contain eggs.

The School Museum contains some larval cases of these insects.

Hepialidæ group.—These creatures are to be found boring into live timber and do a good deal of damage to many forest trees in India. Like other wood-boring larvæ, they have a firm cuticle and powerful mandible. They can be distinguished from wood-boring larvæ belonging to other orders of insects by the possession of the fleshy abdominal pro-legs which are characteristic of caterpillars. The group includes the families Hepialidæ, Cossidæ, and Ægeriidæ. The moths are very variable in their appearance, but the caterpillars are so closely related to each other, both in habits and in appearance, that it will be most convenient for the students to consider them collectively.

In each case eggs are laid by the female upon the bark and the caterpillar tunnels into the wood. It usually keeps the tunnel open and often comes out to feed upon the bark. It frequently builds for itself a covered gallery of silk and excrement which forms a conspicuous feature on tree trunks. The chrysalis is formed at the entrance of the burrow, so that the moth can readily make its exit.

In Travancore, a few years ago, very considerable injury was done to teak trees by a large red boring caterpillar of this group. It was found to commence its attack on the wounds caused by lopping, and its effect was to let the moisture penetrate into the stem and induce rotting. Sandal-wood, coffee, and tea all suffer in a somewhat similar manner from the attack of an allied caterpillar of the species *Zeuzera coffeæ*, Nietner. Young poplar trees also in Baluchistan have been reported as injured to a very large extent by the species *Sphecia ommatiæformis*, Moore. The moth of the last species can be immediately recognised by its striking resemblance to a large wasp. The caterpillars tunnel into the poplar stems close to the ground and weaken the trees to such an extent that they are frequently blown over by the wind. Wood-boring caterpillars can be readily killed by squirting a little kerosine-oil into the hole which they make in the trunk, or by running a sharpened wire into it, but this treatment is only likely to be of use in the case of valuable fruit trees.

The School Museum contains larvæ and moths of this group. The students should notice particularly both the pro-legs and the comparatively firmness of the body in the larvæ.

Bombycidæ group.—The larvæ of most lepidopterous insects are able to spin a certain amount of silk, but there are in India a number of species belonging to the families of Bombycidæ and Saturniidæ which shelter themselves during their pupal stage in cocoons made of large quantities of silk, which can be utilised for commercial purposes. The only forms with which the students need be acquainted are, the domesticated Mulberry silk-worm, the domesticated *Eri* silk-worm, and the semi-domesticated *Tusser* and *Muga* silk-worms. They should notice, however, that there are many wild silk-worms in India which would be useful if they were forthcoming in larger quantities than is now the case.

An account of the structure of the mulberry silk-worm (*Bombyx mori*) is given in the chapter on the dissection of this insect. The imago is an insignificant-looking white moth, with rudimentary mouth parts. Several varieties are cultivated in the plains of Bengal upon mulberry leaves. Most of them differ from the mulberry silk-worm of Europe in being smaller in size and in passing through a number of generations in the year. The large annual variety of Europe, however, is cultivated in Kashmir and elsewhere in the hills. Mulberry silk-worms yield a large amount of valuable silk in India, but are of little importance to the forester. The students should notice that they suffer from the attack of Tachinid flies, and are subject to *pebrine flacherie* and other diseases which likewise attack the European variety.

The Eri silk-worm (*Attacus ricini*, Boisd.) is somewhat larger than the mulberry silk-worm. It is reared in a state of domestication, upon castor-oil leaves in Assam, much in the way that the mulberry silk-worm is reared in Bengal. It passes through a number of generations in the year, and produces a coarse but useful silk known as Assam silk. The moth is a large dark-coloured insect, which may be recognised by its peculiar shape and markings. Eri silk-worms suffer to some extent from the same diseases and parasites as the mulberry silk-worm.

From a forester's point of view the most important silk insects are the *Tusser* and the *Muga*. The *Tusser* silk-worm (*Antheræa mylitta* or *A. paphia*) is a brilliant green caterpillar, adorned with numerous tubercles. The imago is a large yellow or buff-coloured moth, which can be recognised by its shape and by the curious transparent ocelli with which its wings are furnished. The cocoon is a firm structure which is attached to the food plant by a silken stalk of singular strength and neatness. The insect is reared in the jungles of the Central Provinces, Chota Nagpur, and elsewhere throughout the Central and Southern parts of India. The caterpillars feed on the leaves of saj (*Terminalia tomentosa*), sâl (*Shorea robusta*) and ber (*Zizyphus jujuba*), and other forest trees. The insect hybernates in the cocoon. The moths of the first generation emerge in the early part of the rains, and lay eggs which usually produce cocoons about the middle of the wet season. The moths emerge, and lay eggs almost immediately, the cocoons being formed in the beginning of the cold season. Fresh cocoons for breeding are generally collected each March in the jungle. when the foliage is thin. They are usually tied onto the branches of trees, which are pollarded to receive them. The whole life of the insect is thus spent in the open and the trees have to be carefully watched, while rearing is going on, to keep off birds, bats, lizards, and predaceous and parasitic insects, which would otherwise destroy the silk-worms. At best the business is a precarious one. It is still carried on upon a considerable scale, but seems to be upon the decrease.

The *Muga* silk-worm (*Antheræa assama*) is reared in Assam much in the way that the Tusser is reared in Chota Nagpur, though it is somewhat more domesticated. It has much resemblance to the Tusser, but differs both in the nature of the cocoon, which has no peduncle, and in the shape and markings of the moth and caterpillar. It also goes through a series of about five generations in the year, instead of only two, as in the case of the Tusser. According to Stack it is chiefly reared on the sum tree (*Machilus odoratissima*) and on the sualu tree (*Tetranthera monopetala*), but will also feed on the leaves of the mezankuri (*Tetranthera polyantha*) and the champa (*Michelia sp.*). The silk is much like *tusser* in consistency, but is remarkable for its beautiful golden sheen. The students should carefully examine the various specimens illustrative of this group in the School Museum.

Noctues.—This group comprises a great number of moths, which can usually be recognised by their smooth thick-set bodies and filiform antennæ. The larvæ are smooth grub-like caterpillars with the full complement of legs and prolegs. Most of the species are leaf-feeders, but some have a limited power of boring. The chrysalis is usually, but not by any means always, formed in an earthen cell in the ground. The following forms may be noticed :—

(1) The silk cotton-pod caterpillar (*Mudaria cornifrons*, Moore), which destroys the pods of the silk-cotton tree (*Bombax sp.*). The chrysalis is formed in the ground in the beginning of the hot season, and remains there until the early spring of the following year, when the silk-cotton tree puts forth its flowers.

(2) The *cut-worms* belonging to several different species, which attack young plants of all kinds, including coffee bushes, and often do a good deal of damage. The caterpillars are earth-coloured grubs, which live in burrows in the ground. They sally out at night and cut off young plants, which they drag to their holes and feed on at leisure. Dressing

the land with such substances as soot and gas-lime has been recommended for getting rid of them.

(3) *Hyblæa puera*, Cram. (= *Tortrix tectonæ*, Nisbet).—This is a little dusky moth, with reddish markings in the hind wings. The moth is much like a large Tortrix (Microlepidoptera), but has hitherto been included by entomologists amongst the *Noctues*. The caterpillar defoliates teak and has done a large amount of damage in Lower Burma.[1] It has also been reported from Assam, Berar and the North-West Provinces, so is likely to occur in all parts of India. The caterpillar is said to appear in April and May, shortly after the flushing of the spring leaves. About the end of May the caterpillars begin to let themselves down to the ground by silken threads. The pupa is formed in a flimsy cocoon, which is often attached to dead leaves on the ground. Major Bingham notices that a plantation attacked by this insect is a wonderful scene of activity, as numbers of jungle-fowl, ground-thrushes (*Pitta*) and insectivorous birds of all kinds crowd to the spot to feed upon the caterpillars. Mr. Nisbet suggests cutting out of trees that are badly attacked and driving pigs into the forest in the first half of June when the chrysalides are on the ground.

(4) The boll-worm (*Heliothis armigera*), which attacks agricultural crops of all kinds and tunnels into the pods of poppy cotton and leguminosæ. The caterpillar of this insect has little to distinguish it from other Noctues larvæ, but the moth can be recognised by its greyish colour and the arrangement of the dusky streak on the posterior wing.

The students should examine the specimens of Noctues insects in the School Museum and sketch a few typical representatives.

Geometres.—These moths can usually be recognised by their slender build, large wings, and antennæ pectinated in the male. The caterpillars may be known by their peculiar mode of progression. They have the anterior pairs of prolegs rudimentary and the remainder set far back, while the body is slender and elongated. Owing to this peculiar

[1] Mr. J. Nisbet, Deputy Conservator of Forests, writing to the *Indian Forester* on the injury done by this insect in the Pegu Circle, observes:—"These ravages of the *Tortrix Tectonæ* ?) are really of very considerable importance. From April till July, when the summer flush of leaves comes, the increment in growth is lost almost entirely, and the later increment of the year must to some, and in all probability to a considerable extent, have been affected by the denudation of leaves previously. The occurrence of these attacks and details as to their magnitude do not yet find a place in our Annual Reports, although in all likelihood they do as much (or more ?) harm than a fire, unless the latter passes through a plantation where the accumulation of débris is great."

structure they are able to step along with considerable rapidity by holding the ground first with the jointed legs at the anterior end of the body and then with the prolegs at the posterior end, while they alternately hump up their bodies and stretch themselves out to their full extent. Geometres caterpillars are popularly known as *loopers.* They feed upon leaves and are likely to cause damage as defoliators, though they do not seem as yet to have been reported in this connection in Indian forests. The students should notice the group, as the insects are common and conspicuous.

The Museum contains both larvæ and moths for examination.

Microlepidoptera.—This group is a somewhat artificial one, as it is made to include a vast assemblage of minute moths of very different structure. It may be roughly defined as comprising minute moths which do not belong to any of the preceding groups. Amongst the more injurious species the students should notice—

(1) The toon-borer (*Magiria robusta,* Moore), which can usually be seen at work on the young toon trees which line the roads in Dehra. The caterpillar tunnels into the terminal shoots of the toon. Each shoot that is attacked dies back, and as the process goes on, year after year, the trees get converted into masses of distorted branches with no well-marked trunk above a few feet from the ground. The imago is a little greyish moth which emerges in the autumn from the chrysalis which is formed in the tunnel.

(2) The teak-leaf roller (*Paliga damastesalis,* Walker), which has been reported as doing a good deal of damage in Lower Burma, also in Berar. The moth is a pearly-white creature, with numerous tiny spots and cross-bands upon the wings. The pupa is formed on the leaf, which is rolled up by the caterpillar for the purpose. Moths have been obtained in May and the insect is likely to pass through several generations in the year. Major Bingham writes—

> "My recollection (for I unfortunately have mislaid the notes I took) of the ravages of these caterpillars when I was in charge of the Rangoon Division, is that they appeared in the teak plantations on or about the 20th May, sometimes in almost incredible numbers. Their ravages were confined to certain areas where they reappeared year after year, stripping the young teak of their leaves with the rapidity almost of locusts, and hanging in thousands by webs to the branches of the trees."

(3) The numerous clothes-moths, whose caterpillars may be recognised by the little cases that they construct for themselves out of the woollen material upon which they feed. In 1889 some of these insects destroyed the lining of

nearly all the saddles belonging to the Debra Dun volunteers. A plentiful use of spirits of turpentine upon woollen material that is put away is about the simplest way of preserving it from attack, but the insect is a very difficult one to eradicate.

(4) The students should also notice the fact that microlepidopterous caterpillars belonging to various species bore into mangoes, pomeloes, and other fruits, also into the stalks of sugarcane, sorghum, rice, and wheat, besides injuring agricultural plants in other ways. They should also examine the specimens of Microlepideptera in the School Museum and make a sketch of a branch attacked by the toon-borer.

(5).—Chordata.

This group may be defined as comprising bilaterally symmetrical cœlomate animals, which have the body supported by an internal median rod (notochord), which separates the main nerve system from the digestive organs. It comprises the *Urochorda*, the *Cephalochorda*, the *Cyclostomi*, and the *Vertebrata*.

(*a*) Urochorda.—This group comprises the *Acidians* and allied animals, which have the notochord confined to the tail and usually present only in larval life. They are degenerate sea-living animals of small size and are of no practical importance to the forester, though they should be noticed, as they are to some extent intermediate in structure between the Vertebrata and the lower groups of Cœlomata. They breathe water by means of gill-slits, which perforate the pharynx, as in the higher water-breathing members of the Chordata, while the arrangement of the nervous system in the larva is essentially similar to that of the higher forms. The adult animals are degenerate, usually stationary, barrel-shaped organisms. They have the body enclosed in a curious case, whence has arisen the name *Tunicata*, by which the group is often known.

(*b*) Cephalochorda.—This group comprises the *Amphioxus*, which is a small fish-like animal about a couple of inches long, found in shallow seas (*e.g.*, some parts of the Bay of Bengal). It is of no practical importance to the forester, but should be carefully noticed, as it very nearly represents what is believed to be the primitive ancestral form of the vertebrata. The group may be defined as Chordata with notochord persistent throughout life, but without skull, lower jaw, or paired limbs. Both in structure and in development the *Amphioxus* is of the very greatest interest, as it helps us to understand how the Vertebrata came to be what they are; but the subject is one which is too complicated to enter upon in this elementary work.

(*c*) Cyclostomi.—This group comprises the hags and lampreys, some of which are of considerable size. They are mostly more or less marine, some of the species being parasitic. They are of no practical importance to the Indian forester, but should be noticed, as they comprise a very well-marked section of the Chordata. They may be defined as eel-like Chordata, with persistent notochord and skull, but without lower jaw or paired fins.

(*d*) Vertebrata.—This group comprises all the fish, amphibia, reptiles, birds and mammals, and is hence of the utmost importance. The chief points in which the Vertebrata differ from the other groups of Chordata are in possessing a lower jaw and paired limbs, and in the modification of the notochord into a vertebral column. The various sections of the group will be considered under the headings of Fish, Amphibia, Reptiles, Birds and Mammals.

Fish.

Fish are cold-blooded aquatic Vertebrates, with paired pectoral and pelvic fins (corresponding to the fore and hind limbs of the higher Vertebrata). They breathe by gills throughout life. For an account of the anatomy of a typical fish, see the chapter on the dissection of the Mahseer.

Omitting the *Ganoids*, which are of no immediate importance, the chief groups of fishes are the Chondropterygii (cartilaginous fishes), the Teleostei (bony fishes), and the Dipnoi (amphibious fishes).

(*a*) Chondropterygii.—These fish breathe exclusively by gills and have the muscles supported by a cartilaginous skeleton. They comprise the sharks, dogfish, sawfish and skates and rays, which are very abundant round the coasts of India. They are largely carnivorous, some species being dangerous to human life. They are also of some slight importance economically, as the livers of many of the sharks are used for making oil, the skins are made into the shagreen employed for covering knife handles and sword scabbards, while glue is made out of the fins of many species, and the flesh of most of the species is used for food.

(*b*) Teleostei.—These fish breathe exclusively by gills, and have the muscles supported by a bony skeleton. They comprise all the common food-fishes of India. The following are a few of the more important forms:—

I.—*Cyprinidæ* or carps of all kinds, with narrow mouths often furnished with barbels, jaws weak and without teeth, teeth, however, being numerously present on the bones of the throat. The students should at least examine and sketch some such Carp as the Mahseer, if they have not time to do the complete dissection described on pages 89 to 93.

(1) The *Mahseer* (*Barbus tor*) of hill-streams. This fish mounts the streams in the rainy season and lays its eggs, a few at a time, in the small nullahs. In the hot weather it descends into the deeper pools. It feeds on weakly fish, water-snails, fresh-water crustacea, insects and seeds. There are also numerous small carps (genus *Barbus*) allied to the Mahseer, to be found in rivers throughout India. The School Museum contains numerous specimens.

(2) The *Indian trout* (*Barilius bola*), found in rivers with clear streams and stony beds throughout India. A beautifully spotted fish. There are also a number of allied fish (genus *Barilus*), known as the "lesser Barils," which are found in fresh water throughout India. The School Museum contains a representation of *Barilius bola* for examination by the students.

(3) *Chela* or *Chilwa* (*Chela argentea*) and numerous allied species of fly-taking fish, very common in fresh water throughout India.

(4) *Labeos*, including the *Rohita* of Bengal, numerous species of heavy sluggish fish found in tanks and other still fresh water throughout India. The School Museum contains one of these well known tank fish. The students should notice particularly the large suctorial lips with which it is furnished.

II.—*Siluridæ*, known as fresh-water sharks, cat-fish, or butter-fish with broad depressed head and naked skin. They are found in fresh water throughout

India and are very destructive to young fish. Many of them are excellent eating. The Museum contains numerous representatives.

III.—*Fresh-water Eels.*—Numerous species of smooth elongated fishes found in fresh and brackish water throughout India. Many of them descend to the sea to lay their eggs. They are said to be destructive to fish ova.

IV.—*Murrel.*—Several species of almost amphibious fishes, which are found in tanks, wells, rivers and other fresh water throughout India. They have the head covered with plate-like scales and are very destructive to young fish. They often have holes in the bank where they lie. The Museum contains numerous representatives.

V.—*Mugilidæ* (Mullet).—Found in rivers and estuaries, often far above tidal influence. They are common in rice-fields. Some of them swim like tadpoles with their eyes just above the water and are said to be good eating.

VI.—*Spined Eels.*—Elongated fish with sharp spines on the back, found in quiet fresh-water pools throughout India. They are said to be destructive to fish ova. The Museum contains representatives.

VII.—*The Gar-fish.*—With enormously produced beak-like jaws. The Museum contains a specimen of this fish.

Besides the above, which are all fresh-water fishes, there are in India numerous sea and estuary fishes which are largely used as food. The students will not be expected to know much about them, but should notice that the following are the most important forms.

Herrings (*Clupidæ*), such as the *hilsa*, the *oil-sardine* and the so-called *white mullet.* The *hilsa* is a sea-fish, which ascends the rivers of India, especially the Ganges, in vast numbers in the rainy season to spawn, and is excellent food until it has done so, after which it becomes thin and unwholesome. The *oil-sardine* is a smaller fish, which is often very abundant off the Malabar Coast, where it is used for making oil. The so-called *white mullet* is a very active fish, to be found in salt and brackish water on the West Coast, where it is much esteemed for food. *Scopelidæ,* such as the *Bombay duck* of the sea and estuaries. Perches (*Percidæ*), such as the *red perch* of Madras and the *bekti* of Calcutta. *Sparidæ,* such as the *black rock cod* of Madras. *Polynemidæ,* such as the *mango-fish*, which ascends the Ganges and other rivers in the hot weather. *Sciænidæ*, such as the *whiting* of the Bay of Bengal, which is excellent eating when freshly cooked, but rapidly becomes insipid after death. *Labyrinthici*, such as *Anibas scandens*, which is almost amphibious.

(*c*) Dipnoi.—These fish are not found in India, so are of no practical importance, but it is useful to notice them, as they have the swimming bladder converted into lung-like organs, which enable them to breathe air as well as water. They are thus intermediate between fish and amphibia. They occur in Australia, Africa, and South America.

AMPHIBIA.

These are cold-blooded Vertebrata, usually with a naked skin. They breathe as larvæ by gill slits, and as adults by lungs. For details of the structure of a typical Amphibian reference should be made to the chapter on the dissection of the Frog. The chief Amphibia to be met with in India are the bull-frogs, little brown frogs, tree-frogs, and toads. In their adult condition these animals feed on insects and probably destroy a good many destructive species, but they are of no very great economic importance. The chief point to remember about the Amphibia is that they connect the fish with the reptiles and birds on one side and with the mammals on the other.

REPTILES.

The reptiles comprise a number of cold-blooded air-breathing vertebrates, of which the most important are the Chelonia (turtles and tortoises), the Lacertilia (lizards), the Ophidia (snakes), and the Crocodilia (crocodiles). Reptiles resemble birds in having the mandible articulated to the skull by a pair of quadrate bones, and in having the skull articulated to the vertebral column by a single occipital condyle. Students should verify these points by comparing the skull of a reptile with that of a bird. They should also notice the difference in each case from the skull of a mammal where there is no quadrate bone visible but two occipital condyles. Reptiles are cold-blooded; they are covered with scales; the heart consists of two auricles and a ventricle, the latter being more or less completely divided into a right and left chamber. Two aortic arches are present, but the right arch is in many cases the larger and more developed, thus indicating an advance in the direction of the arrangement found in birds, where the right aortic arch alone persists.

(*a*) Chelonia (Turtles and Tortoises).—Body compressed, covered with a bony dermal skeleton, the dorsal portion uniting with the spines and ribs of the vertebral column, the ventral plates uniting to form a covering for the abdomen. The head, limbs and tail can usually be retracted within the dermal armour. The jaws are covered with horny epidermis, which takes the place of teeth. The group comprises the *Tortoises,* which have a hard shell and feet more or less suitable for walking on shore, and the *Turtles,* which have the shell covered with soft skin and the feet modified more or less completely for swimming. Students should examine the specimens in the Museum collection of both turtles and tortoises.

(*b*) **Crocodilia.**—Body partly covered with bony scutes, a long swimming tail, teeth implanted in sockets, *e.g.*, Fish-eating gavials with long snouts, carrion-eating muggers, with short snouts both represented in the School collection.

(*c*) **Lacertilia (Lizards).**—Body covered with scales. There are generally two pairs of limbs, but one pair and sometimes both pairs may be absent. Eyelids present. Jaws firmly united in the middle line. Shoulder-girdle always present. Include water-lizards, geckos, cheecha, and chameleon. Lizards are chiefly insectivorous and destroy a good many injurious species, especially Acridida.

The students should compare the scales of the lizard with the bony scutes of the crocodile, also the movable eyelid of the lizard with the immovable glassy membrane which covers the eye of the snake. The Museum contains numerous specimens.

(*d*) **Ophidia (Snakes).**—Body covered with scales, generally with a series of specially modified scales beneath the body, by means of which the snake glides. Fore-limbs and shoulder-girdle always absent. Hind limbs occasionally present in a rudimentary state (Pythons). Bones of the jaws movable on each other. Eyelids transparent and permanently closed. Includes all the snakes.

Classification of snakes.[1]

The most important families of snakes are the Boidæ (Pythons), the Colubridæ (Colubrine snakes), and the Viperidæ (Viperine snakes).

(*a*) *Boidæ* (Pythons or rock-snakes).—These snakes are often of immense size. They have long oval heads, with numerous small shields (loreals) between the ocular and nasal

(1) The students should work carefully over this section, comparing the specimens in the School Museum, as it is important for them to be able to recognise a poisonous snake when they come across one.

shields. They have rudimentary hind limbs, which terminate in claw-like spurs on either side of the anus. The family does not contain any poisonous species. Students should examine the skin in the School Museum and sketch the arrangement of the shields upon the head, specially noticing the numerous loreal shields.

(*b*) *Colubridæ* (Colubrine snakes).—This family comprises the bulk of the snakes to be found in India. Colubrine snakes have oval heads, but differ from the Pythons in having no recognisable rudiments of hind limbs. They may be classified as (1) harmless land colubrines, (2) poisonous land colubrines, (3) sea colubrines.

(1) *Harmless land colubrines.*—There are a great number of these snakes, many of them being extremely common. They possess numerous small teeth, but the anterior maxillary teeth are neither enlarged nor grooved. The great majority of the species have at least one loreal shield. The little green grass-snake (Dryophis) and the rat-snake (dhamin) may be taken as examples. Students should examine the school collection of harmless colubrine snakes, and sketch the teeth and the arrangement of the scales on the side of the head, specially noticing the loreal shield.

(2) *Poisonous land colubrines.*—These snakes have the anterior maxillary teeth enlarged and grooved for injecting the poison fluid. Practically speaking, they never have a loreal shield. They include a number of deadly snakes, of which the most important are the cobra, the krait, the hamadryas, the raj samp or banded krait, and the Callophis and Adeniophis snakes. The students should examine the poisonous land colubrine snakes in the School Museum and be able to recognise the cobra, the krait, and the hamadryas. They should make a sketch of the teeth, also of the side of the head, to show the absence of the loreal shield.

(3) *Sea colubrines.*—These snakes can be at once recognised by their tail, which is flattened for swimming. They are said to be extremely poisonous, but, being exclusively sea creatures, are of little importance to the forester.

(c) *Viperidæ* (Viperine snakes).—These are poisonous snakes, with triangular heads. They have the anterior maxillary teeth enlarged and grooved for injecting the poison fluid. They can be easily recognised by the peculiar broad triangular head, narrow neck, short tail and the numerous small scales with which the whole body, and especially the neck and the sides of the head are covered. They may be classified as (1) Viperinæ (True Vipers), and (2) Crotalinæ (Pit Vipers).

(1) *Viperinæ* (True Vipers).—These vipers have no loreal pit between the eye and the nostril, and are excessively poisonous. The only species which the students are likely to meet with in India are the Daboia (Chain or Russell's viper) and the Echis. The first can be recognised by its chain-like markings, and the second by its peculiar serrated scales. The students should sketch the head of these forms in the School Museum, noticing particularly the absence of the loreal pit.

(2) *Crotalinæ* (Pit Vipers).—These vipers have a well-marked loreal pit, which is a little hole in the side of the head between the eye and the nostril. They comprise a number of species, which are all poisonous but not so deadly as the true vipers. The students should sketch the side of the head of one of the Crotalinæ from the school collection, indicating the position of the loreal pit, which is the easiest feature for recognising them.

Birds.

Like the reptiles, Birds are air-breathing vertebrates in which the skull articulates to the vertebral column by means of a single occipital condyle. The mandible articulates to the skull by means of a pair of quadrate bones. Birds themselves are warm-blooded; they are clothed with feathers, which are modifications of the epidermis. They have the fore-limbs converted into wings. The jaws are covered with a horny epidermal sheath. The heart is divided into four chambers, and the aortic arch is single and curves over to the right side. The lungs are firmly fixed to the back of the thorax and have air-sacks in connection. For details of the structure of a typical bird, see the chapter on the dissection of the sparrow.

Birds are chiefly of importance in Indian forests—(*a*) as distributors of the seeds of such trees as the figs, mulberry, sandal and the parasitic

Loranthaceæ; (*b*) as the destroyers of seeds, fruit and insect; (*c*) as game; and (*d*) as sought after for plumage.

Birds are classified as *Saururæ*, *Ratitæ*, and *Carinatæ*. The *Saururæ* comprise the fossil species *Archæopteryx*, which differed from all other birds in the great relative length of the tail region of the spine and in the imperfect specialisation of the fore-limbs. The *Ratitæ* comprise such birds as the the ostrich of Africa, the rhea of America, the emu of Australia, and the cassuary of New Guinea. They are incapable of flight, and have no ridge to the breast-bone for the attachment of the wing muscles. The *Carinatæ* comprise the remaining birds, including all the species that occur in India. They are characterised by the possession of a ridge to the breastbone for the attachment of the wing muscles.

The Carinatæ comprise an enormous assemblage of species, which are most difficult to classify on account of the great number of intermediate forms between the different groups. The chief features that are usually taken for classificating purposes are the nature and arrangement of the bones in the roof of the mouth[1] and the arrangement of the tendons[2] by which the toes are flexed. These features, however, are not only somewhat obscure, but the conclusions to be deduced from them are so indefinite that hardly any two writers upon the subject have hitherto adopted precisely the same classification. In his *Birds of Burma* Mr. Oates, who is the author of the first half of the work upon the Birds in the *Fauna of British India*, classifies the Carinatæ into the groups Passeres, Macrochires, Pici, Coccyges, Psittaci, Striges Accipitres, Steganopodes, Herodiones, Anseres, Columbæ, Gallinæ, Geranomorphæ, Limicolæ, Gaviæ, Turbinares, and Pygopodes. This arrangement is somewhat artificial, and the characters upon which it is based are too complicated for the students to attempt to master, but it is about the most convenient that is at present available, so may be adopted provisionally. The Turbinares (diving petrels) and the Pygopodes (grebes) may be passed over as of no practical importance to the forester.

I.—Passeres (Perching birds).—This group comprises an enormous gathering of species, of which the house-crow and the sparrow may be taken as typical examples. The students should go over the named collection of birds in the School Museum and learn to recognise at least the house and jungle crows, tree-pies, bulbuls, mynas, and sparrows, which feed upon a mixed diet, and the king-crows, wagtails, babblers and swallows, which are insectivorous. A sketch of the head and beak, with a note on the size of the bird and the colours of its feathers, will usually be

[1] Especially the vomer and the maxillopallatine bones, also the nasal and frontal bones.

[2] The *flexor perforans digitorum* and the *flxeor longus hallucis*.

sufficient to enable the student to remember any particular species with which he may happen to be unfamiliar.

II.—Macrochires.—This group includes the Swifts and Goat-suckers, both of which feed on insects. They may be recognised by their wide mouths and weak feet, which generally have all the toes directed forward. Students should be able to recognise the Swifts and Goat-suckers as such in the Museum collection.

III.—Pici (Wood-peckers).—These birds have feet with big claws formed for climbing. One of the toes is often very much smaller than the others. They have a long pointed tongue for seizing the insects on which they feed. The students can easily recognise them by their stiff tail feathers, which help to support them on the trunks of trees as they peck the insects out of the dead-wood.

IV.—Coccyges.—This is an ill-defined group, which comprises a number of very different families. The students should go over the Museum collection and be able to recognise the green barbet, the coppersmith, the koel, the hoopce, the green bee-eater, the Indian roller, and the kingfishers as such. Of these the koel, the hoopoe, the green bee-eater and the Indian roller are insectivorous. The green barbet and the coppersmith feed on a mixed diet, while the kingfishers are aquatic feeders. The Indian roller and the kingfishers are often killed for their plumage.

V.—Psittaci (Parrots).—These birds can be easily recognised by their peculiar hooked beaks and weak feet. They have the upper jaw movably hinged on to the forehead. They comprise all the green parrots, which are excessively destructive to grain and fruit in India. They are killed to some extent for the sake of their plumage. The Museum contains numerous specimens.

VI.—Striges (Owls).—These birds can be recognised by their powerful hooked beak and claws and the circlet of stiff feathers that surround the eye. They are mostly nocturnal in their habits and feed upon small animals of all kinds. The School Museum contains numerous examples.

VII.—Accipitres.—These birds have beaks and claws very much like the owls, but can at once be distinguished from the owls by the absence of the circlet of stiff feathers around the eye. They include all the Vultures, Eagles and Hawks, which feed on flesh and carrion. The Museum contains numerous examples.

VIII.—Steganopodes.—This group comprises the pelicans, cormorants, and snake-birds, with webbed feet and comparatively short legs. They are all large in size. The pelicans may be recognised by the bag-like expansion beneath the lower jaw, while the peculiar hooked beak of the cormorant, and the slender neck and pointed beak of the snake-bird are very characteristic features of these creatures. The Steganopodes all feed on fish and other aquatic food and are of no great practical importance. The pelicans are said sometimes to do a little damage by destroying fish in jheels, while the snake-bird is killed for its plumage.

IX.—Herodiones.—These are wading birds, with long legs and long beak. They live about water and feed chiefly on frogs. They include the ajutant (stork), herons, and egrets. Egrets, such as the cattle-egret and paddy-bird, are largely killed in India for their plumage, but the group is not otherwise of much importance. The Museum contains a specimen of a paddy bird.

X.—Anseres (Ducks and geese).—The birds of this group can readily be recognised by their broad beaks covered with soft richly innervated skin, and their webbed feet for swimming. The students should be able to recognise the ducks, geese, and teal as such. These birds are all largly used for food. The Museum contains some of the ducks.

XI.—Columbæ (Pigeons and doves).—These birds can be at once recognised by the soft weak beak swollen at the base. The students should be able to recognise the green wood-pigeon, the blue rock-pigeon and the common spotted ring-doves as such in the collection. The Columbæ feed on grain and seeds and are valuable as food.

XII.—Gallinæ (Game birds).—These are large thick-set land birds, with powerful beak and feet for scratching up the soil. They feed on a mixed diet of seeds, insects and shoots, and are valuable for food. Some of the pheasants and jungle-fowls are also sought after for their plumage. The students should be able to recognise the pheasants, partridges, jungle-fowl, spur-fowl, pea-fowl, sand-grouse and quail, as such, in the School Museum.

XIII.—Geranomorphæ.—This group includes the bustards, florikins, cranes, and rails, which mostly have long legs and comparatively short beaks (*e.g.*, the large sarus crane with red head, which is said to be used as food). The

Geranomorphæ feed on a mixed diet of fruit, grain and insects and are of no great importance.

XIV.—Limicolæ.—This group includes the plovers, snipe, curlews and stints, which have feet modified for walking on soft mud and often a long beak. The plovers and snipe are valuable for food, and the students should be able to recognise them in the School Museum.

XV.—Gaviæ.—This group is very closely related to the Limicolæ; it comprises the gulls, river terns and skimmers, which are none of them of any practical importance.

With regard to the time of the year during which nesting goes on in India, Mr. W. L. Sclater writes—"Most small birds, in Upper India at any rate, breed between April and July. Of course there are many exceptions; but the four months—April, May, June, and July—would practically cover the breeding time of all the birds which require protection. In Southern India many birds breed in December and January, and in the hills the breeding season, as, for instance, in the case of the monaul, is in July and August. In the case, however, of Lower Bengal, the best months are undoubtedly April, May, and June." (Extract from *Indian Museum Notes*, volume II, No. 5.)

MAMMALS.

Air-breathing warm-blooded vertebrates, more or less covered with hair. They nourish their young for a time after birth with the secretion of mammary glands. The skull articulates to the vertebral column by two occipital condyles. The aorta is single and bends over to the left side. The heart consists of four completely divided chambers. For an account of the main features in the anatomy of a typical mammal, see the chapter on the dissection of the palm squirrel. The students should compare the skull of a bird with that of a mammal to notice the single condyle in the one and the two condyles in the other, also the apparent absence of the quadrate bone in the mammal. They should also sketch the teeth in the skull of some such mammal as the jackal, so as to learn the arrangement of the incisors, canines, premolars and molars.

Mammals are usually classified into the three groups—*Prototheria, Metatheria* and *Eutheria*. The *Prototheria* comprise the duck-billed platypus and the spiny Echidna, both of the Australian region. These creatures are of no practical importance to the Indian forester, but they are interesting as representing the lowest existing types of mammalia. They are characterised by the generalised nature of the brain and bones, the simplicity of the generative and urinary organs, and by the fact that unlike other mammalia, they lay eggs which contain a nutritive yolk as in the case of birds. The *Metatheria* include the kangaroos and wombats of Australia and the opossums of South America. They are to a great extent intermediate in structure between the ***Prototheria*** and the ***Eutheria.*** The young are born at a very early stage of development and are then nourished in a pouch which contains the mammary glands. Like the ***Prototheria,*** the ***Metatheria*** are of no practical importance to the Indian forester, but should be noticed as representing a large and interesting section of the class. The ***Eutheria*** comprises all the mammals that occur in India. They are characterised by a special arrangement (placenta) by means of which the fœtus is nourished within the uterus of the mother, and by the possession of a distinct urogenital opening.

The Eutheria are classified into the orders Edentata, Sirenia, Cetacea, Carnivora, Ungulata, Rodentia, Insectivora, Chiroptera, and Primates.

I.—Edentata.—These are lowly mammals, with imperfectly developed teeth. The only representative that need be noticed is the Indian Manis or pangolin, which can at once be recognised by the large horny scales with which the upper part of the body is covered. The pangolin burrows into the earth and feeds on ants and Termites (white-ants). The students should be able to recognise the pangolins' skins as such in the School Museum.

II.—Sirenia.—These are aquatic vegetable-feeding mammalia which are of no practical importance to the forester. They comprise the dugong, which is to be found on the coasts of India. The hind limbs are absent and the front limbs, neck and head are partially modified for aquatic life. The mammæ are pectoral.

III.—Cetacea.—This order comprises the whales of the Indian ocean, also the dolphins and porpoises, some of which are found in the tidal rivers of India. They are fish-like mammalia, without hind limbs, with the front limbs converted into fins, and the whole body far more specialised for aquatic life than the Sirenia. The mammæ are inguinal.

IV.—Ungulata.—This order comprises all the hoofed mammals, which are very numerous and important. The Indian representatives are all included in the sub-orders Proboscidia, Perissodactyla, and Artiodactyla.

(a) *Proboscidia.*—With five toes, includes the Indian elephant. *N. B.*—The trunk is merely a modification of the nose, while the tusks are modifications of the upper incisors. The grinding teeth (molars), of which there are six pairs, are pushed forward successively from behind and shed in front as they wear out. The Museum contains a skull in which the students can examine the teeth.

(*b*) *Perissodactyla.*—With less than five toes, the third (middle) digit being the largest. Includes the horse, donkey and wild ass of Sind, with foot consisting of but one digit, the second and fourth digits being rudimentary and forming splint-bones on either side of the middle toe; also the rhinoceros, with three toes, and the tapirs, with four toes in front and three toes behind.

(*c*) *Artiodactyla.*—With two median digits (third and fourth) equal in size. This sub-order includes a large number of Indian species:—(1) Pigs, which have a simple stomach and are non-ruminant; (2) Camels, which are ruminants, with incisors in both jaws; (3) Musk-deer, which are ruminants with enormous canines, but no incisors in the upper jaw; (4) Barking deer (*Cervulus mountjac*), ruminant, incisors absent above, canines large, horns on long bony stalks; (5) Sambar, hog-deer (*Cervus porcinus*), Swamp deer, and Chital—all ruminants, canines small or absent, with solid dermal horns, which are cast off at intervals, no incisors in the upper jaw; (6) Mouse-deer, without horns, ruminant—but with only three compartments to the stomach instead of four as in the true deer, with canines, but without incisors in the upper jaw; (7) Common antelope, four-horned antelope, gazelle, domestic goat, nilgai, serow, ibex, tahr, burrhal, oorial, ammon, Ovis poli, gaur (Indian bison), yak, buffalo, and Indian cattle,—all ruminants, with neither canines nor incisors in the upper jaw, the horns hollow and supported by a bony centre. The students should go over the representatives of the Artiodactyla in the School Museum and be able to recognise them. They should pay particular attention to the teeth and the horns.

With regard to the relative damage done by various Ungulates in Indian forests, goats are probably about the most destructive, as they continually nibble off the young shoots and thus prevent seedlings from growing into trees. Cattle and sheep again do an immense amount of injury of a similar nature, though to a somewhat lesser extent. Elephants break down young trees, eat the leaves of several species and also strip the branches of their bark. In the hot weather they are said to be particularly partial to sâl bark, while bamboo clumps are often completely uprooted by them. Deer do some damage by browsing on the tops of seedlings and coppice shoots, also by rubbing their horns against young trees. Pigs are said to devour great quantities of acorns and other fallen seed. They also uproot seedlings, and sometimes injure the bark of trees by rubbing their tusks on the trunk. On the other hand, their grubbing up of the soil is said to be an excellent preparation for the reception of seed, specially in deodar forests, where the main obstacle to reproduction is the thick matting of fallen needles, on which the seed is liable to dry up without germinating. In coniferous forests again pigs are said to destroy the larvæ and pupæ of injurious species of moths which hybernate beneath the litter. Upon the whole, Mr. Clifford inclines to the belief that in hill forests pigs are more useful than injurious, while in the plains the reverse is the case.

V.—Rodentia.—With two large chisel-shaped incisors in front of each jaw, which continue to grow throughout the life of the animal. For details of structure, see the chapter on the dissection of the palm squirrel. The Rodents include the squirrels, bandicoot rats, black rats, brown rats, mice, porcupines and hares. The students should be able to recognise the squirrels, rats, porcupines, and hares as such in the School Museum. Squirrels and rats do some damage by devouring seed, while porcupines often injure seedlings by gnawing the bark.

VI.—Insectivora.—The representatives of this order may be recognised by the large number of teeth in both jaws, the molars being furnished with sharp points for crushing insects. Include hedgehogs and musk rats, which destroy a good many insects, but are of no very great importance to the Indian forester.

VII—Chiroptera.—This order comprises the bats of all kinds, which have the fore-limbs specialised for purposes of flight. Bats are mostly insectivorous and destroy a good many injurious insects; a few species, of which the large flying fox (Pteropus) is the best known, feed upon fruit and are likely to be of some importance in distributing seeds.

VIII—Carnivora.—These mammals have the teeth, and in many cases the claws specialised for tearing flesh, one of the premolars (carnassial) being often modified for cutting. The more important species which occur in India belong to the families Ursidæ, Mustelidæ, Canidæ, Hyænidæ, Viverridæ, and Felidæ. The students should go over the specimens of carnivora (especially skulls) in the School Museum and be able to recognise the groups given below :—

(*a*) The *Ursidæ* comprise the bears, which have plantigrade feet, and two molars above and three below, with broad tuberculated crowns.

(*b*) The *Mustelidæ* contain the martens, weasels, badgers and otters. They have but one true molar behind the carnassial tooth above. The body is long and the limbs short. Five toes usually present, both in front and behind.

(*c*) The *Canidæ* contain dogs, wolves, jackals, and foxes. They are digitigrade carnivora, with non-retractile claws. The wild sppcies to be found in India all have four toes behind and five in front, the pollex (thumb) being much shorter than the other digits and not reaching the ground. There are four premolars and two molars above.[1]

(*d*) The *Hyænidæ* contain the common striped hyæna, which is at once recognisable by the prominence of the bony crest in front of the head and the relative shortness of the hind legs. It has non-retractile claws, but the teeth approach those of the cats in the small development of the molars, of which there is but one in the upper jaw.

(*e*) The *Viverridæ* contain the civets, paradoxures (tree-civets), mungooses and their allies. They are closely related to the Felidæ, but differ from this family in their elongated heads and more numerous back teeth. They are distinguished from the *Mustelidæ* in having the auditory bulla of the skull rounded in shape.

(*f*) The *Felidæ* contain all the cats, including the tiger, lion, leopard, cheeta, lynx, and snow leopard. They are distinguished from other carnivora by

[1] Jackals are omnivorous and do a good deal of damage by feeding on sugarcane stalks. They also eat coffee berries and aid very largely in the distribution of this plant, for the seed passes through their digestive organs without injury. The fact is said to be so well known that in parts of Madras the right to collect what is known as "Jackal coffee" in the forest is let out on lease.

their short rounded heads and highly specialised canine and carnassial teeth. They never have more than three premolars followed by one molar above. The claws are completely retractile and the muscular system is extraordinarily developed.

IX.—Primates.—These mammals may most easily be distinguished by the fact that they have the eyes enclosed in distinct sockets. They are divided into the *Lemuroidea,* with eye-sockets open behind, and the *Anthropoidea,* with eye-sockets closed behind. The *Lemuroidea* include the lemurs, which are of no practical importance to the forester. The *Anthropoidea* include all the monkeys and man. The chief families of monkeys to be found in India are—

(1) the *Cercopithecidæ,* in which a tail is almost always present and which include the common langurs and macacus monkeys;

(2) the *Simiidæ,* in which no tail is developed, and which include the howling hoolock (gibbon) monkeys of the jungles of Assam and Burma.

PRACTICAL COURSE.

INTRODUCTORY.

Each student should bring with him to class a pair of pointed scissors, a pointed pen-knife or scalpel, a pair of forceps and a magnifying lens. He will be lent a piece of weighted bark, a dissecting dish and pins from the School stores. These appliances are all that are necessary for the dissections that are described in the following pages, but any men who happen to have additional dissecting instruments should bring them. A couple of mounted needles can be very easily made by fixing big darning needles into little pieces of wood for handles. They are a specially useful addition. So also is a thin steel knitting needle with rounded end, to use as a probe. The knitting needle should be heated in the fire and then allowed to cool slowly, as this enables it to be bent to the shape required. The scissors and knife should both be as sharp as possible. In the case of the scissors it is important that the blades should meet up to the ends when closed. Sharp-pointed nail scissors will do very well if regular dissecting scissors are not available.

EXAMINATION OF PROTOZOA.

In Dehra the Protozoa, or one-celled animals, which are most easily procured belong to the genera (1) Paramecium, (2) Opalina and (3) Vorticella.

Parameria are to be found swimming freely in stagnant pools everywhere. *Vorticellæ* can usually be found by keeping a bottle of stagnant water from one of the rice-fields for a few days, until it becomes somewhat foul. *Opalinæ* occur in large numbers in the large intestine of the common Dehra frog. To obtain living specimens of *Opalinæ* for examination, chloroform a frog and as soon as it is dead, open up the abdominal cavity and snip out the large intestine with a pair of scissors. Then slit up the side of the piece of intestine and wash out the contents with a little salt solution in a watch glass. The mixture thus obtained will usually be found to be swarming with *Opalinæ*.[1]

In the case of each form, a drop of water containing the animals should be put upon a slide, covered with a slip of thin glass, and examined with an inch objective to find a satisfactory specimen. After a specimen has been found, a quarter-inch objective may be used with advantage.

The students not being individually provided with microscopes, the examination of these Protozoa must be confined for the present to what can be made out in the few minutes during which each member of the class can have one of the available microscopes to himself. After looking down the microscope, each man should make a sketch of what he has been able to make out.

Paramecium appears as a flattened semi-transparent slipper-shaped animalcule, covered with vibratile hairs (cilia), by means of which it swims rapidly through the water. On its lower surface is a large groove which leads to the mouth. Its movements are so rapid that the students will have some difficulty in making out its structure, which is that of a one-celled animal with nucleus and contractile vacuoles. Waste products are cast off through an aperture (anus) which is situated between the mouth and the hinder end of the body. Reproduction is effected by fission, which is sometimes preceded by the temporary joining up (conjugation) of two individuals.

Opalina is very much larger and less active, so it is far easier to make out. It appears as a white oval creature covered with large cilia, by means of which it swims. It has neither mouth nor anus, as it feeds entirely by absorbing fluid from the digestive tract of the frog. It has

[1] Postscript—In 1893 no *Opalinæ* could be found to show to the students though in the preceding year the parasite simply swarmed in the digestive tract of every frog that was examined.

numerous nuclei, but no contractile vacuoles. It reproduces itself by simple fission.

Vorticella is a transparent bell-shaped animacule. It is attached to a long stalk, by which it anchors itself. From time to time the students will see it draw itself down by coiling up the stalk like a corkscrew and then suddenly shooting itself out so as to stretch the stalk straight. Round the margin of the bell the students will be able to see a fringe of large cilia, which vibrate rapidly. Both the mouth and the anus lie in a ciliated groove, which occupies very much the same place as the end of the clapper would occupy in a bell. There is a nucleus and contractile vacuole, but the students will probably not have time to make these out satisfactorily. *Vorticella* reproduces itself both by fission, and also by spore-formation preceded by the permanent conjugation of a small free swimming individual with a larger stalked one.

The white blood corpuscle.

A good example of an amœboid cell is the white blood corpuscle. Chloroform a frog, extract a drop of its blood and spread it on a slide under a cover slip. Examine with the highest power available of the microscope. Amongst masses of flat oval yellowish discs, which are the red blood corpuscles, the students will be able to make out a few colourless granular cells, which, when carefully watched, can be seen to extrude finger-like processes or pseudopodia.

DISSECTION OF THE EARTH-WORM. [1]

Earth-worms (*Lumbricus sp.*) are common in damp places about Dehra, where they are known as *kenchua.* The largest available specimens should be procured for dissection. They can be readily killed either by dropping them into alcohol, or with chloroform. Examine a specimen with a lens and make out the following features :—

(1) The numerous little transverse grooves which divide the body into rings or segments.

(2) The thickened reddish section of the body (cingulum), which occupies several segments in the anterior third of the length.

(3) The delicate chitinous membrane (or cuticle) with which the whole body is invested.

(4) The four pairs of short spines (setæ) borne on the ventral surface of most of the segments. They can be felt on drawing the worm through the fingers.

[1] The common species of earth-worms in India differ markedly from each other in the arrangement of their reproductive organs. The following account therefore, which has been drawn up from some specimens found near Calcutta in December, will not always apply completely in this particular. This is of little importance, however, as the essential features of the worm's structure are the same in each case.

(5) The mouth, which is the anterior opening of the digestive tract. It is situated at the front end of the body.

(6) The anus, or posterior opening of the digestive tract, which is situated at the end of the last segment.

(7) The genital openings, which lie on the ventral surface. The students will best be able to make them out when they have dissected the organs with which they are connected.

(8) The mid-dorsal pores, which are minute holes that serve to place the body cavity in communication with the exterior. They lie between the segments in the mid-dorsal line.

Stretch the anterior segments of the worm a little, by gently pulling the animal through the fingers.

Pin the animal, ventral surface downwards, on to the weighted bark. Two pins will be sufficient. One of them should be inserted through the lip and the other as far back as possible. Now with scissors make a longitudinal slit in the dorsal wall of the integument. Work from behind forwards, and take great care not to cut through more than the outer layer. Dissect back the flap on either side, cutting through the numerous little membranous partitions by which the digestive tract is attached to the outer wall. Pin back the flap on either side. The students will now have exposed the following organs :—

(1) In the centre line a longitudinal vessel (dorsal vessel) which lies immediately above the digestive tract, and can at once be recognised by the red colour of the blood it contains.

(2) The digestive tract, which is a thick, soft, dark-coloured tube throughout the greater part of its length.

(3) The blood-vessels, which pass round the digestive tract and connect the dorsal vessel with the ventral vessel which lies below.

(4) The reproductive organs, which lie on either side of the digestive tract.

Now examine the digestive tract in detail and observe that it is a straight tube which extends from the mouth in front to the anus behind. Make out the following parts in it, examining from in front backwards :—

(1) The mouth.

(2) The pharynx, with expanded muscular walls.

(3) The œsophagus, which connects the pharynx with the gizzard.

(4) The gizzard. This is an expanded whitish organ, with thick muscular walls, in which the food is crushed before passing into the soft intestine.

(5) The long straight intestine, with dark-coloured bulging walls. It extends throughout the rest of the body and connects the gizzard with the anus.

Now examine with a lens the blood-vessels which pass round the digestive tract from the dorsal vessel. Notice that five of the vessels which encircle the intestine, immediately behind the stomach, are much expanded. These have contractile walls and are sometimes called hearts. They serve to keep up the circulation of the blood, which courses up the dorsal vessel and down the ventral vessel.

Now examine the reproductive organs. These vary a good deal in different species and at different times of the year, but the students will usually be able to make out the following parts :—

(1) A large pair of white globular sacs (receptacula seminis), in which the spermatozoa, received from another individual, are stored. These sacs lie on either side of the œsophagus and open by slits to the exterior below.

(In one common Calcutta species there are three of these sac-like recepticula, one behind another.)

(2) A pair of large yellowish glandular organs (vesiculæ seminales), which contain the testes. These lie on either side of the intestine behind the stomach and close to the three hindmost pairs of hearts.

(3) A pair of white tubes (vasa deferentia) which are sometimes much coiled; the greater part of them lies some distance behind the vesiculæ seminales. They serve to convey the spermatozoa to the male genital openings, which lie on the ventral surface of the cingulum.

The ovaries are smaller and more difficult to make out, and it is only exceptionally that the students will be able to see them. They lie posteriorily to the vesiculæ seminales, the eggs passing out by a pair of short tubes, which lie one opposite each ovary.

The students should now make a sketch of the dissection and mark in the names of the organs they have been able to recognise.

Carefully dissect out the digestive tract, so as to expose the ventral blood-vessel, which runs down the body immediately below the digestive tract. Trace where the connecting vessels from the dorsal vessel join it, and notice the branches that run along the sides of the œsophagus. Below the central vessel see the large white nerve (ventral nerve chain), swollen slightly in each segment. Trace it throughout the whole length of the animal, noticing the little nerves which are given off from each swollen place (ganglion). Immediately behind the mouth the students will be able to make out that a pair of large nerves are given off. These pass round the digestive just in front of the pharynx and connect the ventral chain with a pair or nerve masses (supra-æsophageal ganglia), which lie above and in front of the œsophagus.

The excretary organs of the earth-worms consist of a number of little tubes (nephridia), which open to the exterior, a pair to each segment. They are very difficult to make out, however, in the Dehra species and the students will not be able to recognise them.

DISSECTION OF THE FRESH-WATER CRAB.

Fresh water-crabs (*Telphusa sp.*) are common under stones in the beds of the Song and other rivers of Dehra Dun. They are known locally as *gegra* or *kenkra*. The largest available specimens should be procured. They can be killed almost instantaneously by dropping them into boiling water. If boiling water is not available, chloroform may be used.

Examine a specimen closely and observe that it is covered with a hard armour (exoskeleton) which serves in all Arthropoda to support the muscles. Make out in it the following parts :—

(1) The body, which consists of—

(*a*) The broad anterior unsegmented portion (cephalothorax) with a curved groove across the dorsal portion of it.

(*b*) The little posterior segmented portion (abdomen), which consists of six divisions folded up under the cephalothorax.

(2) Five pairs of large jointed appendages attached to the ventral surface of the cephalothorax.

(3) The jaws and other small jointed appendages which surround the mouth.

(4) The mouth, which is to be seen by separating the appendages which surround it.

(5) The pair of movable eye-stalks, and the two pairs of minute jointed antennæ, which are situated in front of the cephalothorax.

(6) The anus, or posterior opening of the digestive tract, situated on the lower surface of the last segment of the abdomen.

(7) The little jointed appendages attached to the abdomen.

Notice that the crabs are of two kinds, *viz.* :—

(1) Males, with narrow abdomens, to which are attached two pairs of little white appendages close to the cephalothorax.

Roy. 8vo. Cloth. pp. vii, 242. 10s. 6d.

UMMAGGA YĀTAKA

(The Story of the Tunnel).

TRANSLATED FROM THE SINHALESE

BY

T. B. YATAWARA, M.C.B.R.A.S., Ratemahatmaya.

"There is a mine of information, as well as of unintentional humour, in this narrative of a birth of Buddha in one of his previous lives. It gives us a wonderfully vivid insight into the customs and ways of thinking of mediæval India."—*Saturday Review.*

"This is a unique specimen of Indian literature, of the sort that makes glad the heart of Max Müller, delights the lover of Oriental romance, furnishes much food for thought to students of Eastern classics, and is not without entertainment of a novel kind for English and American readers who have plenty of time at their disposal and like to explore out-of-the-way shelves in the great library of the world and take down volumes that do not get into everybody's hands."—*Literary World.*

LONDON: LUZAC & CO.,

Publishers to the India Office,

46, GREAT RUSSELL STREET, OPPOSITE THE BRITISH MUSEUM.

(2) Females with broad abdomens, to which are attached four pairs of bifurcated swimming appendages.

In the female make out the paired genital openings, which are easy to see on the ventral portion of the cephalothorax, opposite the insertions of the last pair but two of large ambulatory legs.

In the male look for the genital openings, which are situated at the base of the last pair of legs. They are much more difficult to make out than the genital openings of the female.

Lift up the jointed appendages which surround the mouth and notice on either side the anterior opening into the branchial chamber. Also lift up the whole cephalothoracic shield as far as it will go without tearing the membranes underneath it, and see the posterior openings above the last pair of ambulatory legs, through which the water enters the branchial chamber.

Now make a sketch of the crab, both from above and from below and mark in the names of what has been seen.

With a pointed pair of scissors cut away one side of the shield of the cephalothorax very carefully, above the bases of the ambulatory legs, beginning at the anterior opening of the branchial chamber. This will expose the large branchial chamber. Notice the numerous branchiæ which occupy the chamber; they are attached just above the bases of the legs.

Examine one of the branchiæ with a lens and see how it consists of a central portion to which are attached a number of delicate processes arranged like the leaves of a book.

With a pair of scissors cut away the edges of the shield which covers the cephalothorax. Dissect off the whole shield, a little piece at a time, so as not to damage the organs which lie below it. In doing this be very careful not to tear any of the tissues, for the heart lies immediately below and is liable to be torn away with the muscles by which the stomach is attached to the centre of the shield.

If small pin the animal, dorsal surface upwards, on to the weighted piece of bark, inserting the pins into the bases of the legs. If large it will be better not to pin it down. Now immerse the crab in water deep enough just to cover it.

Notice that the whole space which has been exposed by dissecting off the cephalothoracic shield is covered with a delicate dark-coloured membrane. Dissect off the whole of this membrane very carefully under water, removing any thread-like parasites (Nematodes) which you come across.

The following organs will now be visible:—

(1) On either side the large branchial chamber, which we have already examined from below.

(2) The large muscular stomach, which lies in the middle line of the body immediately above the mouth. It is attached to the shield by powerful muscles, a pair of which are visible in front.

(3) The pericardial chamber. This is a transparent thin-walled sack lying in the middle line immediately behind the stomach. Its function is to receive the blood from the gills and to pour it into the heart. It contains the heart, from which proceed the thin-walled arteries which convey the blood to all parts of the body.

(4). The straight intestine, which can be seen underneath the heart, passing from the stomach towards the abdomen.

(5) The liver, this is a soft mass of little yellowish lobes lying on either side of the stomach and ramifying between the other organs throughout the greater part of the cephalothorax. It pours its secretions into the digestive tract by ducts which lie behind the stomach.

In the female see the ovaries, which in November appear as bright yellow masses filled with eggs and protruding on each side above the liver. Attached to the ovaries are a pair of wide thin-walled ducts (oviducts), which you can trace on either side passing down towards the genital openings on the ventral surface.

In the male, in place of the ovaries, see a pair of soft white elongated glands (testes), which pass up on either side between the heart and the stomach. Endeavour to trace the ducts which pass downwards on either side of the heart to connect the testes with the genital openings at the base of the legs.

Carefully remove the whole of the stomach by cutting through the intestine behind, the muscles in front and the gullet below. Cut it open, wash out the contents, and see the three powerful teeth it contains. Observe how these teeth work against each other for crushing the food.

Wash out the cavity in the cephalothorax from which the stomach has been taken, and cut out as much of the intestine as is visible, so as to see how the liver is connected with it. Cut away the reproductive organs and any lobes of the liver that remain, taking care not to cut away the tissues which underlie these organs.

Immediately in front of where the stomach lay, and close to the outer wall of the cephalothorax, the students will now be able to make out two soft curved white glands which open immediately in front of the mouth. These probably correspond with the green glands of the cray-fish, in which case they are excretory organs.

Look for the small white supra-œsophageal nerve ganglion, which lies in front of the gullet, and trace the nerves which pass back from it around the gullet to join the ganglia, which lie on the ventral floor of the body.

Now unpin the animal, turn it over and examine the appendages, pulling them off one by one with forceps, beginning from behind, and sketch them in order in a note-book. The appendages are as follows, working from behind forwards :—

(1) Attached to the abdomen :—

(*a*) In the female, four pairs of bifurcated swimming appendages.

(*b*) In the male, two pairs of small pointed appendages.

(2) Attached to the cephalothorax :—

(*a*) Four pairs of large jointed legs on which the animal walks.

(*b*) One pair of large jointed pincers (chelæ). The pincers are formed by the last joint, which works against a prominence on the last joint but one.

(*c*) Three pairs of jointed appendages (maxillipedes), the most posterior of which forms a pair of broad plates which cover up the mouth.

(*d*) Two pairs of little white appendages (*maxillæ*), the posterior pair of which are modified into oval plates (scaphognathites), which serve to pump the water out of the branchial chamber.

(*e*) One pair of hard jaws (mandibles), each with a little jointed appendage (palp).

(*f*) One pair of movable stalked eyes.

(*g*) Two pairs of minute jointed antennæ.

Notice that the little swimming appendages, attached to the abdomen of the female present the same essential structure as all other appendages.

Notice the little auditory organs at the base of the front pair of antennæ.

DISSECTION OF THE COCKROACH.

The best cockroaches for dissection belong to the species *Periplaneta americana*, Deg. This is a large form which is scarce in Dehra Dun, but easily procurable in Calcutta. A large number should be procured. They should be preserved in alcohol until wanted for dissection. The fresher they are the better. In Calcutta they are known as *tilchatta*.

Amongst the specimens procured for dissection there are pretty certain to be some immature ones (larvæ). Notice that they are wingless, but otherwise very much like the full-grown individuals.

Examine a full-grown specimen carefully with the help of a hand lens. Notice that the whole animal is covered with plates of a firm, hard substance (chitin), which encases the body and limbs like armour. Notice the joints, where the plates are replaced by soft skin, so that the animal is able to bend its body and limbs freely. This armour is found in all arthropods; it serves at once to protect the body and to support the muscles.

Make out the three regions of head, thorax, and abdomen. The head is the small box-like portion in front which bears the mouth, eyes, and long jointed feelers (antennæ). The thorax is the middle portion of the body. It is covered by three rings of plates, which divide it into three well-marked portions (segments), which can easily be made out. The front segment (prothorax) is covered above by a broad plate with yellow markings on it. Below it bears a pair of long jointed legs. The second segment (mesothorax) is smaller; to its upper (dorsal) surface are attached the front pair of wings, while to its lower (ventral) surface are attached the middle pair of legs. The third segment (metathorax) is very similar to the mesothorax; to its dorsal surface are attached the hind pair of wings, while to its ventral surface are attached the third (last) pair of jointed legs.

The abdomen consists of a number of segments (the students will be able to count about nine of them), telescoped into each other. The last segment bears a pair of jointed appendages (anal cerci) attached to its sides.

Notice that the specimens are of two kinds:—

(1) Males, with a little pair of styles attached to the ventral surface of what looks like the eighth, but what is really the ninth, segment of the abdomen.

(2) Females, without anal styles, but with a large genital pouch, with slit-like orifice, on the ventral surface of the hinder segments of the abdomen.

Examine the legs and notice that they are very similar to each other in structure, each consisting of the following joints:—

(1) A large basal joint (coxa).

(2) A minute second joint (trochanter).

(3) A third joint (femur), to which are attached a few short bristles.

(4) A fourth joint (tibia), with numerous large bristles.

(5) A series of small joints (tarsus), with a pair of terminal claws. The last tarsal joint (pulvillus), which bears the claws, is minute.

Examine the wings and notice that they consist of thin membrane, supported by numerous nervules.

Examine the appendages which surround the mouth *in situ* with a lens, so as to see how they lie with regard to each other. Then pull them out carefully one by one with forceps, beginning from behind, and notice that they consist of the following :—

(1) A jointed bilobed flap (labium), with a three-jointed appendage (palp) attached to each side of it.

(2) A pair of broad bilobed appendages (maxillæ), each with a five-jointed appendage (palp) attached to its side.

(3) A pair of hard-toothed jaws (mandibles.)

(4) A movable flap (labrum), which closes the mouth in front.

(5) A fleshy tongue (lingua) which is an out-growth of the front portion of the digestive tract.

Examine the spiracles, or openings of the breathing organs, on either side of the body. There are two large ones on each side between the bases of the legs, which the students may be able to make out with the help of a lens, if the specimen is fresh. There are also eight pairs of stigmata situated on the sides of the abdomen between the margins of the plates; these are more difficult to make out.

Carefully cut off the wings close to the base, and notice the dark longitudinal streak which extends up the centre of the back of the thorax and abdomen. The streak indicates the position of the dorsal vessel, or heart, which lies immediately underneath the armour plates.

Now pin the specimen, ventral surface downwards, to the weighted bark, inserting the pins into the bases of the legs, so as not to interfere with the dissection of the body. Immerse in water deep enough just to cover the specimen. Dissect off the whole of the dorsal plates, taking great care not to cut into the organs lying below.

On cutting off the plates a layer of whitish fat will be exposed, with a dark longitudinal space running up the middle line. The dark longitudinal space contains the dorsal vessel, but as the walls of this organ are thin and transparent, and the blood is colourless, it is somewhat difficult to make it out.

The dorsal vessel is a straight many-chambered tube which receives the blood from the cavity of the body through openings on either side, and by its rythmical contractions drives it forward towards the head.

Now dissect off the fat carefully so as to expose the digestive tract, which should be uncoiled and separated from the fat and net-work of tracheal tubes that surround it. Trace it up to the throat in front, taking care not to cut away the soft white salivary glands. Then clear it down to the posterior end of the abdomen. Notice that the digestive tract is a tube, extending from the mouth in front to the anus behind, and that it consists of the following parts :—

(1) The mouth, into which the two pairs of diffuse white salivary glands pour their secretion.

(2) A narrow gullet.

(3) A large crop, which occupies the greater part of the body cavity.

(4) A little muscular gizzard which is armed inside with six little brown teeth something like what we found in the fresh-water crab, only on a smaller scale.

(5) A narrow thin-walled portion (mesenteron), into the front part of which a series of large glandular tubes (hepatic cæca) open.

(6) A narrow convoluted portion (proctodœum), which connects the mesenteron in front with the anus behind. Notice the numerous thread-like secretory organs (malpighian tubes) which pour their secretion into the digestive tract, just when the mesenteron joins the proctodœum.

Notice the little silvery tubes (tracheæ) which ramify throughout the whole body, carrying air from the spiracles at the sides. In some lights the students will be able to make out with a lens the spiral filament by which the walls of the tracheæ are supported, but usually a microscope is required for this purpose.

Now carefully brush away the fat at the posterior end of the abdomen. If the specimen is a female, make out the two sets of slender tubes (ovaries), in which can be seen a number of white thickenings, which are the eggs. If the specimen is a male, make out the large white tufts of glandular tubes (mushroom gland), which receive the spermatozoa derived from the pair of minute testes, which the students will not be able to make out.

In the centre line of the abdomen, close to the ventral armour, the students will now be able to make out the ventral nerve system. This consists of a series of little white nerve masses (ganglia), each connected with the next by two nerves, so as to form a chain which extends throughout the whole length of the body. Trace this chain as far as possible in both directions. Notice the little nerves given off from each ganglion, and in dissecting the portion about the neck, try to make out the nerves which pass round the gullet, and connect the nerve mass which lies in the head (supra-œsophageal ganglion) with the front ganglion of the ventral chain. Now make a careful drawing of the dissection and mark in the names of the organs that have been successfully made out.

DISSECTION OF THE MULBERRY SILK CATERPILLAR.

Mulberry silk caterpillars (*Bombyx*) were for a long time cultivated in Majra in the Eastern Dun. They can now be procured either from Cashmere, where the large species *Bombyx mori* is to be had, or from Lower Bengal, where only smaller varieties are obtainable. The largest available full-grown specimens should be procured. They can be killed by dropping them in spirit, where they can be preserved until required for dissection, but the fresher they are the better. If it is necessary to preserve them for any length of time in alcohol before dissection, each specimen should be pricked in several places with a needle to help the spirit to make its way into the tissues.

Examine a specimen carefully with a lens and notice that it consists of the following parts:—

(1) A round head, which bears the minute antennæ, eyes and mouth parts.

(2) A thorax, consisting of three segments (prothorax, mesothorax, and metathorax), each of which bears a pair of jointed legs. These are the true legs, corresponding to the legs of the cockroach.

(3) An abdomen, in which the students will be able to make out about nine segments. The two anterior segments are without limbs. The four following segments (Nos. 3 to 6 inclusive, counting from before backwards) each have a pair of fleshy prolegs attached to them. Behind these are two segments (Nos. 7 and 8) without limbs. No. 8 has a small mid-dorsal spine. Last of all is a collection of somewhat indistinct segments bearing the last pair of fleshy prolegs.

Notice the nine pairs of button-like openings (spiracles) which are connected with the breathing tubes (tracheæ) that ramify through the body. They are situated on each side of the body. The anterior pair lies on the prothorax, while each of the anterior eight segments of the abdomen bears one pair.

Notice a faint longitudinal streak along the middle of the back. This is due to the dorsal vessel (heart), which lies immediately below the integument. It is liable to be missed in rough dissection. In a living specimen its rythmical pulsation can easily be seen through the intervening tissues. Its function is to pump the blood towards the head.

Now make a sketch in a note-book indicating what has been seen and marking each part with its name.

Pin the caterpillar, ventral surface downwards, on to the weighted bark, inserting the pins into the anal prolegs and the sides of the thorax, so as not to interfere with the internal organs. Immerse the bark in water deep enough just to cover the specimen. Carefully open up the body cavity by cutting down the mid-dorsal line with a pair of pointed scissors. Dissect back the flap on either side, taking great care not to injure the organs which lie below. Pin back the flaps, dissect off the loose yellowish fat which is usually present in large quantities, and the following organs will be exposed :—

(1) A thick straight digestive tract extending down the middle of the body, from the mouth in front to the anus, which lies in the last segment of the abdomen between the anal prolegs.

(2) A delicate dorsal vessel or heart, which lies in the middle line above the digestive tract. This will very probably have been destroyed in opening up the cavity.

(3) The silk glands. These are two long, firm, much convoluted organs which lie on either side of the digestive tract. They contain a fluid which hardens into silk when exposed to the air.

(4) The numerous silvery trachea connected with the stigmata on either side of the body.

Notice that the digestive tract is a tube opening at each end of the body, and that it consists of the following parts :—

(1) The cavity of the mouth.

(2) A narrow gullet.

(3) A large dilated portion (stomach), which extends throughout the greater part of the body.

(4) A narrow intestine into which open three pairs of slender thread-like excretory glands (malpighian tubes).

(5) An expanded chamber (rectum) terminating in a posterior opening (anus).

Dissect out the silk glands and trace the duct on either side which passes up into the mouth.

Make a rough sketch in a note-book showing the organs exposed, cut off a small piece of one of the largest tracheal tubes that can be found and endeavour with a lens to make out the spiral filament by which it is kept in shape.

Cut away the digestive tract at the gullet, pin it back out of the way. Pin the silk glands also out of the way on either side, and search for the chain of brilliantly white nerve ganglia which lie along the ventral surface of the body beneath the digestive tract. When this has been found, dissect away the tissues around it so as to expose it throughout its whole length. Notice the numerous nerves given off from each ganglion, and endeavour to make out the branches which pass round the gullet to connect the supra-œsophageal ganglion, which lies in the head, with the sub-œsophageal ganglion, which lies below the gullet, and is the first of the ventral chain.

DISSECTION OF THE SNAIL.[1]

Large snails are not easily obtainable in Dehra, though small ones are common enough. There is a large land species (*Achatina fulica*, Ferussac) which is common in Calcutta and which does very well for dissection. It is known locally as *ghonga*. The snails should be killed by drowning them, as by this means they are obtained in an expanded condition. They should be allowed to remain in the water for about two days, after which they can be preserved in alcohol until wanted for dissection.

Examine a specimen and notice—

(1) The coiled shell set on the animal's back, with its apex pointing to the right side.

(2) The expanded fleshy foot upon which the snail progresses.

(3) The head in front, which bears two pairs of tentacles. The large upper tentacles each terminate in an eye, which can only be seen when the tentacle is fully expanded.

(4) The mouth, which is a large median opening in front of the head, closed by large fleshy lips.

(5) The pedal gland, which opens in a large median hole beneath the mouth. Stick a match into the opening of the pedal gland and feel how far it goes back into the substance of the foot.

(6) The genital opening. This is a small round hole which lies on the right side, a little behind and below the upper tentacle. From it the ova and spermatozoa are extruded at different times of the year.

(7) The large opening of the pulmonary chamber. It lies on the right side, just where the shell meets the back. Stick a match into it and feel the extent of the large pulmonary chamber which lies inside the shell.

Cut and break away the shell carefully so as not to injure the animal. Begin from just above the pulmonary opening, and extract the animal completely. Notice the transparent membrane which forms the upper wall of the large pulmonary chamber. Stick a thin match into the opening of the pulmonary chamber, and by feeling about in the right hand upper corner of it you will be able to make out the anus and posterior portion of the digestive tube, which opens in this spot.

Immerse the specimen in water and arrange the coil as it was when inside the shell. Just above where the pulmonary chamber terminates, on the left of the body, is a rounded chamber with transparent walls (pericardium) which contains the heart. To the right of it lies the large yellowish kidney, which occupies the space between the heart and the

[1] It may be necessary to curtail or omit this dissection for want of time.

posterior end of the digestive tract. Above this is the greenish mass of the liver, which extends up to the top of the shell. Below notice the large (columella) muscle by the contraction of which the animal draws itself back into its shell.

Pin the specimen, ventral surface downwards, on to the weighted bark. Open up the pulmonary chamber with scissors and dissect off the membrane with which it is covered. Notice the numerous vessels containing colourless blood, which ramify through the walls of the breathing chamber. See the large pulmonary vein with which the branching veins unite. The pulmonary vein conveys the blood back to the heart after it has been oxidized in the pulmonary chamber.

Slit open the pericardial chamber and observe the heart which it contains. The heart consists of a thin-walled auricle, which receives the blood from the pulmonary chamber, and a thick-walled ventricle, which pumps the blood to all parts of the body, whence it finds its way back to the pulmonary chamber. The kidney is in communication with the pericardial chamber; its duct (the ureter) passes down to open close to the anus.

Find the anus and push a piece of match or grass up into the digestive tract, which is thin walled and sometimes rather difficult to see through the investing membranes. Then with scissors carefully dissect away the membrane around so as to clear the digestive tract. Trace it through its convolutions in the mass of the liver and dissect it out throughout its entire length until it terminates in the mouth in front. In doing this take care not to cut through the fibrous collar by which the digestive tract is encircled just behind the mouth. Notice that the digestive tract consists of the following parts :—

(1) A large mouth in front, which contains the toothed tongue (odontophore) by means of which the snail gnaws the leaves on which it feeds.

(2) A narrow gullet (œsophagus), which connects the mouth with the expanded crop behind. It passes through the thick fibrous collar, which contains the chief nerve centres.

(3) A thin-walled, expanded crop, to which a quantity of white glandular matter (salivary gland) is closely attached. The salivary gland pours its secretion through slender ducts which open into the mouth.

(4) A thick-walled, expanded loop or stomach, which receives the ducts from the liver.

(5) A narrow, thin-walled, winding intestine, which traverses the liver, and ultimately passes into the rectum, by which it is connected with the anus.

On the right-hand side of the digestive tract the complicated reproductive organs will now be exposed. Dissect them out by snipping away the connecting membranes, and make out the following parts. They are very complicated, and the students will not be expected to have more than a general idea of their structure:—

(1) The small irregular glandular body (hermaphrodite gland) which is embedded in the upper coils of the liver. In this the ova and the spermatozoa are alike said to be developed.

(2) A slender, much convoluted duct (hermaphrodite duct), which connects the hermaphrodite gland with the expanded portion below.

(3) The large, firm, white albumen gland, which lies at the side where the hermaphrodite duct terminates.

(4) The large common duct, which is imperfectly divided longitudinally into two passages, one for the ova and the other for the spermatozoa. Sometimes one side of the common duct will be found to be crammed with yellowish-white eggs.

(5) Beyond the common duct, the passages for the exit of the ova and spermatozoa separate into an oviduct and vas deferens respectively. Follow these down to the common opening, noticing the long sac-like branch of the oviduct (receptaculum seminis) in which the spermatozoa received from another individual are stored.

Notice the large columella (retractor) muscle, which sends branches to the tentacles, mouth, reproductive organs, etc. It is attached, as we have seen, to the shell, and serves to retract the various organs.

Dissect away the fibrous tissue which composes the outer part of the broad collar which surrounds the œsophagus, and make out the pair of yellowish-white nerve masses (supra-œsophageal ganglia) which it contains. Trace the two nerves which pass down on either side from the supra-œsophageal ganglia, to join the large nerve mass (sub-œsophageal ganglia) which lies below, thus constituting a complete ring round the gullet. The large nerve masses above and below the gullet give off nerves which supply the sense organs and other parts of the body. In the substance of the sub-œsophageal nerve mass is imbedded a pair of small auditory organs which are connected by slender nerves with the supra-œsophageal ganglia.

Slit open the mouth and extract the toothed tongue-like organ (odontophore) which occupies a large sac in the lower part of the mouth. Examine it with your lens and make out the teeth.

DISSECTION OF THE MAHSEER.[1]

The Mahseer (*Barbus tor*), which is exceedingly common in the streams about Dehra, may conveniently be taken to illustrate the more

[1] It may be necessary to curtail or omit this dissection for want of time.

essential features of structure to be found in fishes. Specimens six or eight inches long are large enough for the students to dissect.

Notice the characteristic flattened spindle-shaped body without neck and with the trunk passing imperceptibly into the tail, also the scales, which cover the body and lie upon each other like roof tiles. Make out the following:—

(1) The large toothless extensile mouth, with enormous fleshy lips and two pairs of barbels. (*N.B.*—Teeth are present in the throat.)

(2) The large eyes, which are without eyelids.

(3) The nose pits, which consist of two little holes on either side situated in front of the eyes. Notice that the nose pits do not communicate with the mouth, and are not used in respiration.

(4) The lateral line. This is a well-marked line running down either side of the body from the head to the tail. It is formed by the openings of a series of little tubes which are believed to be sense organs.

(5) The large gill chamber on either side of the head, covered with a bony plate (Operculum).

(6) The anal and urinary openings, which are situated close to each other in the middle line below, in the posterior third of the body. The anterior opening is the anus.

Locomotion in the water is chiefly effected by the action of the tail, which is supported by powerful muscles, but the fins also assist. The fins of the Mahseer are as follows :—

(1) The dorsal fin. This is the large unpaired fin in the middle of the back. Notice how the front ray is converted into a large stiff spine.

(2) The anal fin. This is the unpaired fin below, just behind the anus.

(3) The paired pectoral fins, which are situated just below the gill opening on either side. They correspond to the front limbs in other vertebrates.

(4) The paired pelvic fins. These are situated on either side of the ventral surface about the middle of the length. They correspond to the hind limbs in other vertebrates.

Make a sketch of the fish from the side, and mark in the names of the structures noticed above.

Lift up the operculum and examine the gills. Observe that they consist of four leaf-like organs, each with a hole (gill slit) on either side of it communicating with the mouth. Notice the little horny processes (gill rakers) between the bony supports of the gills.

(N.B.—*The following description of the internal anatomy is taken from immature specimens of the Rohita fish of Bengal. It will be found to apply in all important points to the Mahseer.*)

Pin the fish on to its side on the weighted bark. Slit the skin up the middle of the side, from the tail to the operculum, with scissors, dissect it away and pin it back. The powerful longitudinal muscles will now be exposed. Notice how they are split up into segments by transverse partitions. Notice also the large white cutaneous nerve which issues from beneath the operculum.

Immerse the specimen in water and cut away the muscle on the ventral surface until the abdominal cavity is found, taking care not to cut into the organs which lie below. Then dissect off the whole of the muscles so as to expose the abdominal cavity completely throughout its entire length. Notice the dark-coloured membrane (peritoneum) with which the abdominal cavity is lined. A fold of this membrane supports the viscera. By moving the viscera about, without further dissection, the following organs can be made out :—

(1) The exceedingly long and slender intestine, the coils of which occupy the greater portion of the ventral portion of the abdominal cavity[1]. In front the intestine passes into a thicker tube (stomach), which is connected with the mouth by means of the gullet. Behind, it communicates with the anus, thus constituting a winding tube from one end of the body to the other.

(2) In the front portion of the abdominal cavity the reddish glandular organ (liver), with large greenish gall bladder, communicating with the digestive tract below the stomach.

(3) Dorsal to the digestive organs, the large transparent swim-bladder, which contains air. In full-grown carps, like the Rohita and Mahseer, the communication between the swim-bladder and the digestive tract is completely closed.

(4) Dorsal to the swim-bladder, the large reddish kidney into which the dark-coloured distended caudal vein passes from behind.

(5) The slender transparent duct (ureter) through which the secretion of the kidney passes out behind the anus. This is a little difficult to see.

Unravel the intestine with the help of fingers and scissors. Sketch the whole dissection. Then cut away the digestive tract and the lower part of the swim-bladder so as to clear the abdominal cavity and enable the kidney to be seen plainly. Dissect away the muscular wall between

[1] In the Mahseer the intestine is shorter and less slender.

the gills on the ventral surface below the mouth, taking care not to injure the blood-vessels which lie in the closed pericardial chamber below. Cut off the operculum on the side which is uppermost and the pericardial chamber will be easily explored. Notice the thin membrane (pericardium) which envelopes the heart, very much in the way that the peritoneum envelopes the viscera.

Make out the following parts of the heart from behind forwards :—

(1) The large thin-walled sack (sinus venosus) into which the blood passes from the liver, kidney and other parts of the body.

(2) The thin-walled vessel (auricle) which lies in front of the sinus venosus.

(3) The thick-walled pinkish organ (ventricle) which lies in front of the auricle and almost completely conceals it.

(4) The little pink thick-walled organ (bulbus arteriosus) which lies in front of the ventricle. It is the commencement of the vessel through which the blood is driven into the branchial arteries.

By pulling back the bulbus arteriosus with forceps and dissecting away the tissues carefully in front, the continuation of this vessel can easily be made out to where it gives off the paired branchial arteries to the gills. From the gills the blood is gathered up by arteries, which unite to form the dorsal aorta. The dorsal aorta is a straight pinkish tube which runs down the back immediately beneath the vertebral column. It can readily be traced by dissecting away the kidney. After oxidisation in the gills the blood passes down the dorsal aorta, which supplies the tail and all the posterior of the body. The blood from the tail returns *viâ* the kidney, where it is partially purified. That from the digestive organs returns *viâ* the liver, where the nutriment absorbed from the digestive tract is elaborated. From both kidney and liver the blood passes into the sinus venosus, whence it is driven by the heart to the gills.

Cut away all the ventral portion of the specimen, leaving only the head and the first inch or two of the back. Pin this portion, dorsal surface upwards, on to the weighted bark, and cut away the back of the head with scissors, so as to expose the brain. This dissection is best performed upon a specimen which has been hardened in alcohol, but most of the points can be made out without much difficulty in the fresh state. The brain is small and lies deep in the skull. After cutting away the bone, pick off with forceps the tissue which surrounds the white nerve-masses of the brain, and make out the following from behind forwards :—

(1) The cerebellum. This is the large median lobe at the posterior of the brain. Behind it can be seen the medulla oblongata, which is little more than a slight expansion of the spinal cord.

(2) In front of the cerebellum a pair of large rounded nerve-masses (optic lobes), which supply the eyes.

(3) In front of the optic lobes a pair of smaller closely appressed nerve-masses (cerebral hemispheres). From the cerebral hemispheres a pair of nerves are given off which communicate with the small olfactory lobes in front. These, however, are very liable to be obliterated in rough dissecting.

Dissect off the muscles down the back and cut away the dorsal spines and arches with scissors, so as to expose the spinal cord. Trace it backwards as far as the specimen admits. Then sketch the dissection.

Lift up the spinal cord carefully with forceps and turn the whole brain forwards, so as to be able to examine its ventral surface. Make out :—

(1) The numerous nerves which leave the ventral surface of the medulla oblongata on either side.

(2) The pair of large optic nerves, which place the eyes in communication with the optic lobes.

(3) The paired bean-shaped nerve masses (lobi inferiores) situated immediately underneath the optic lobes. They enclose between them a small dark-coloured nerve-mass known as the pituitary body.

DISSECTION OF THE FROG.

Frogs are excessively common in Dehra in every pool of water. They can be easily caught with a small hand-net. The largest available specimens should be procured for dissection. If possible, some tadpoles (immature frogs) also should be obtained, to show the gill-breathing stage through which the Amphibia pass. The largest species of frog in Dehra is the Bull-frog (*Rana tigrina*), and the following description therefore is taken from this species. Any of the other forms, however, will do almost equally well, though they are somewhat more difficult to dissect on account of their smaller size. Frogs are known locally as *maindak.*

Any number of frogs can be killed in a few minutes by putting them into an earthen pot (gumla), closing up the top with a dissecting dish and pouring in a little chloroform.

Examine a specimen and notice the blunt head in front, the thickset body behind, and the two pairs of jointed limbs, the whole being clothed with smooth moist skin.

Notice that the specimens are of two kinds :—

(1) Males, with bag-like infoldings of the skin (vocal sacs) on either side of the lower jaw.

(2) Females, without these vocal sacs.

In the head make out the following parts :—

(1) The large wide mouth in front.

(2) The two large eyes situated above the head. Notice the semi-transparent lower eyelid with which the eye can be covered up.

(3) The two nostrils situated between the eyes and the end of the mouth.

(4) The circular smooth tympanic membrane on either side of the head behind the eye.

Open the mouth as wide as you can, and cut down on either side a short way with your scissors, to still further increase the gape. Make out the following :—

(1) The large soft bifid tongue, which occupies the floor of the mouth. It is attached in front and free behind.

(2) The wide œsophagus, into which the mouth passes behind.

(3) The slit of the glottis, through which air passes down into the lungs. It lies in the middle line behind the tongue.

(4) The two posterior nasal openings in the roof of the mouth. They are connected with the nostrils in front.

(5) The openings of the large eustachian recesses, which lie on either side at the back of the roof of the mouth. Each communicates with the cavity behind the tympanic membrane.

(6) The minute teeth with which the upper jaw and roof of the mouth are armed.

Examine the fore-limb and make out—the basal joint or branchium, which is supported by the humerus bone; the second joint or forearm, supported by the fused radius and ulna; and the hand, consisting of four digits, corresponding to the four fingers in man, the thumb being rudimentary.

Examine the hind limb and make out; the basal joint (femur), which is supported by the femoral bone; the leg, which is supported by the fused tibia and fibula; the tarsal joint, supported by two elongated tarsal bones; and the five digits, which are connected together by webbed membrane.

Lay the animal on its back on the weighted bark and pin it down firmly, inserting the pins in the muscles of the fore and hind limbs.

Slit open the skin down the middle line from the lower jaw to the hind legs, noticing the large spaces which lie between the skin and the muscles. Pin back the flap of skin on either side, so as to expose the muscular wall of the abdomen. Notice the large (muscular cutaneous) vein which is usually visible on either side; this is a branch of the subclavian vein, which we shall come to later.

Open up the muscular wall, a little on one side of the middle line, taking great care not to cut into the organs which lie below. Cut through the coracoid bone behind and the clavicle in front, which connect the breast-bone (sternum) with the shoulder joint. By clearing away the muscular wall on either side you will be able to make out;— in the middle line, the sternum, which terminates above and below in wide cartilaginous flaps; and on the side the shoulder joint, where the humerus bone terminates in the glenoidal fossa formed by the junction of the clavicle and the coracoid bones in front and the scapula behind.

Cut out the sternum altogether, taking care not to injure the heart which lies below. Now pin back the flaps of muscular tissue on either side, taking care not to damage the large central vein (anterior abdominal), which lies in the middle line. Notice also the large vein (muscular cutaneous) which leaves the muscle and joins the sub-clavian vein on either side. You will now have exposed—

(1) In the middle line in front, the large heart, invested in a transparent membrane (pericardium).

(2) On either side behind the heart, the large reddish flaps of the liver.

(3) Below the liver, you will see the stomach and part of the intestine coiled up in the lower part of the body cavity.

Move the intestine carefully to one side, without cutting anything, and see the delicate urinary bladder, which lies in the hinder part of the body cavity. Also notice the two dark-reddish elongated kidneys that lie behind, and the yellow finger-like processes of the fat body.

Now dissect off the pericardial membrane, so as to expose the vessels of the heart and make out—

(1) The thick muscular chamber (ventricle), which occupies the apex of the heart.

(2) The distended thin-walled auricles which lie at the base of the ventricle and from the outside appear to consist of but one chamber.

(3) The thick-walled tube (truncus arteriosus) which arises from the right of the base of the ventricle in front of the auricles, and almost immediately divides into a right and left aortic arch.

(4) On lifting up the ventricle and carefully bending the heart forward towards the mouth, you will be able to make out the large thin-walled blood chamber (sinus venosus), where the blood collects before passing into the auricle. The sinus venosus consists of two chambers. The right chamber receives the venous blood from the two venæ cavæ superiores and the vena cava inferior, while the left,

which is much smaller, receives the blood from the pulmonary veins.

Follow down the anterior abdominal vein to the base of the legs, and make out where it branches to each thigh. Then follow it upwards towards the heart and endeavour to make out where it divides into branches, which enter different lobes of the liver. The anterior abdominal vein is one of the two channels by which the blood from the legs can return to the heart. The other being *viâ* the renal portal system.

Cut away the anterior abdominal vein, and bend the liver over forwards with your forceps, noticing the greenish gall bladder. Notice also the venæ portæ, which collect the blood which comes from the stomach and intestine, and pour it into the liver.

Cut away the liver and notice the large vena cava inferior below it, by which the blood returns from the liver and lower part of the body to the sinus venosus. Endeavour to make out the external jugular vein which passes down in front of each aortic arch. Follow this vein towards the heart and endeavour to make out where it is joined by the sub-clavian vein from the front limb and the innominate vein from the back of the head and shoulder. The three together form the superior vena cava, through which the blood returns to the sinus venosus, whence it passes into the right auricle.

Now trace up the aortic branches as far as you can, and endeavour to make out how they divide on either side into three great trunks. The anterior of these (carotid) supplies the head and throat; the middle (systemic aortic arch) runs round the gullet to join the corresponding arch on the other side, on its way giving off the sub-clavian artery to the fore limb; while the posterior (pulmo-cutaneous artery) supplies the lung and gives off a branch near the shoulder to the skin.

Push aside the intestine and trace the inferior vena cava downwards to where it originates in branches from the pair of dark-coloured kidneys, which lie in the posterior part of the body cavity. Push aside the kidneys also and notice the large vein (renal portal) which passes into the kidney from the hind limb. This is the second channel through which the blood from the hind limb is able to return to the heart.

Now cut through the membrane by which the intestine is held together. Push it aside, clear away the vena cava inferior between the kidneys, and below you will find the straight dorsal aorta lying above the vertebræ. This vessel supplies blood to all the lower part of the body. Trace it both forwards and backwards as far as you can. Anteriorily it is formed by the junction of the two systemic aortic arches, while posteriorily it divides into the two iliac arteries, which supply the hind limbs. Between these two points it gives off numerous branches to the digestive tract, reproductive organs, kidneys and spleen.

Clear the digestive tract throughout its entire length, by cutting through the mesentery, noticing the small red spleen, which is attached to the mesentery near the back of the abdominal cavity.

Observe that the digestive tract is a winding tube which extends from the mouth in front to the cloacal aperture behind. Make out the following parts in it:—

(1) The mouth in front.
(2) The soft thin-walled gullet (œsophagus), which connects the mouth with the stomach.
(3) The elongated stomach, with thick glandular walls.
(4) The narrow coiled tube (small intestine) which lies behind the stomach. The anterior end being the duodenum, into which the bile and pancreatic ducts open.
(5) The large intestine, which connects the small intestine with the cloaca.

Endeavour to make out the pale-coloured gland (pancreas) which lies near the commencement of the small intestine.

On either side of the stomach above, observe the large pulmonary air-sacs or lungs. Endeavour to make out the short air-passage (laryngo-tracheal chamber) by which the lungs are connected with the slit-like glottis in the floor of the mouth.

Examine the dark-red elongated kidneys and endeavour to make out the ducts by which their secretion passes to the cloaca.

In the male make out the pair of yellowish rounded testes, which lie in front of the kidneys. Their ducts enter the kidneys and communicate with the ducts of these organs.

In the female notice the broad convoluted ovaries, in which you will usually be able to see numerous little dark-coloured eggs. The eggs pass out by a pair of much convoluted ducts (oviducts) which enter the cloaca.

Cut away the digestive, reproductive, excretory and pulmonary organs, so as to clear the body cavity and expose the spinal nerves. Also cut away the lower jaw. Observe the slender white spinal nerves which issue from the spinal column on either side. In the posterior part of the abdominal cavity these nerves are large and numerous, where they constitute what is known as the sciatic plexus. Follow down these nerves on one side where they enter the leg, cutting away the muscles and bone, and see how they unite to form the large sciatic nerve of the leg. Notice the large brachial nerve which supplies the fore-limb and follow it back to where it issues from the spinal column.

Look for the delicate nerve cords, with swellings (ganglia) at intervals, which connect the roots of the spinal nerves with each other on the same side. These constitute what is known as the sympathetic nerve system.

The more complete dissection of the nervous system can best be performed on a frog which has lain some time in alcohol, as this hardens the nerve tissue. Most of the important features, however, can be made out without this preliminary.

Turn the frog over and dissect off the skin from the back and head.

Cut away the muscles along the vertebral column on either side, and dissect out the flat scapular bones above the shoulders. Find the big opening (foramen magnum) at the base of the skull, where the spinal cord passes into the head. Stick the point of your scissors into it and carefully cut away the top of the head, so as to expose the brain. Great care is necessary in doing this not to injure nerve tissue. When the brain is fairly exposed cut away the vertebral arches in the same manner down the back, so as to expose the spinal cord. Pick away the darkish membrane (pia mater) which covers the white nerve matter, and make out the following parts, beginning from in front :—

(1) Two elongated lobes, each with a constriction in the middle. They lie side by side and occupy the greater part of the length of the brain case. The slender portions in front of the constrictions are the olfactory lobes, while the swollen portions behind the constrictions are the cerebral hemispheres.

(2) Behind the cerebral hemispheres a broad flat portion (thalamencephalon) ; upon the front part of this, between the ends of the cerebral hemispheres, lies the small pineal body, which you may have some difficulty in making out.

(3) Behind the thalamencephalon two large rounded masses situated side by side. These are the optic lobes.

(4) Behind the optic lobes a narrow transverse band of nerve-matter. This is the cerebellum.

(5) Behind the cerebellum is the expanded anterior portion of the spinal cord. It is known as the medulla oblongata.

(6) Behind, the medulla oblongata is continued into the spinal cord. Trace this backwards as far as you can, noticing the roots of the nerves that are given off from it on either side.

Lift up the olfactory lobes in front, breaking through the nerves (olfactory) which they give off to the nasal cavities in front. Notice the large optic nerves which are given off below to the eyes, and behind these, in the middle line, the small prominence which is known as the pituitary body. Pull out the spinal cord as far down as possible, noticing the roots of the nerves which it gives off on either side, and the groove which extends down its ventral surface.

DISSECTION OF THE SPARROW [1].

To make out the essential features of bird structure a sparrow (*Passer domesticus*) may conveniently be taken. Sparrows are excessively common in Dehra. They are best caught alive with nets. Any number can be killed in a few minutes by putting them into a closed earthen *gumla* and pouring in some chloroform. The ordinary native name for the sparrow is *gauriya*. Notice that the body is almost completely covered with feathers. The specimens will be found to be of two kinds:—

(1) Males, with a black patch on the throat and in front of the mouth.

(2) Females, where these markings are replaced by grey, which is far less conspicuous.

In the head examine—

(1) The large laterally-placed eye with its upper and lower movable eye-lids, and the third eye-lid (nictitating membrane), which lies in the anterior angle of the eye and can be rapidly drawn over the eye itself. Pull the nictitating membrane over the eye with your forceps and notice how it works.

(2) The beak, formed by horny sheaths covering the upper and the lower jaw, which are both toothless.

(3) The nostrils, which are a pair of slit-like holes situated on either side at the base of the beak communicating with the mouth behind.

(4) The ears, which appear as circular holes in the sides of the head behind and below the eyes. They are covered with a circlet of small feathers.

Cut down the sides of the mouth with your scissors, so as to increase the extent to which it can be opened, and make out:—

(1) The pointed tongue, which lies in the floor of the mouth.

(2) The large median slit-like opening in the roof of the mouth communicating with the nostrils.

(3) The small median opening immediately behind the large one. This opening is the termination of the eustachian tubes, which place the mouth in communication with the internal cavity of the ear on either side.

(4) The continuation of the mouth behind into the œsophagus, down which the food passes into the crop.

(5) The median slit-like opening which lies behind the tongue and communicates with the windpipe (trachea), through which the air passes into the lungs and air-sacs.

[1] It may be necessary to curtail or omit this dissection for want of time.

Open the wing and feel the various bones by which it is supported. These comprise—

(1) A basal joint (humerus).

(2) A second joint, composed of two bones (*radius* and *ulna*) placed side by side.

(3) A series of bones (carpals and metacarpals) which support the large wing feathers. Notice the first digit or thumb, which supports a small bunch of feathers known as the bastard wing.

In the leg make out—

(1) The basal joint, supported by the femur.

(2) The second joint, supported by a bone known as the tibio-tarsus. This bone is formed by the fusion of the tibia with some of the tarsal bones. It has, closely attached to its outer side, the slender fibula bone.

(3) The third joint, supported by a straight bone known as the tarso-metatarsus or more loosely the tarsus. It is formed by the fusion of the remaining tarsal bones and some of the metatarsals.

(4) The four toes of the foot, the first toe (hallux) being directed backwards.

(5) The claws, in which each toe terminates.

Pull out one of the large feathers from the wing or tail and examine it with your lens to make out the following parts :—

(1) The tubular semi-transparent portion or quill. Notice the two apertures into the tube of the quill, one of considerable size at the end, where the quill penetrates the skin, and the other a minute one, which lies on the ventral surface of the feather just where the quill joins the solid portion (rachis).

(2) The solid shaft (rachis) which is the continuation of the quill.

(3) The flattened lateral expansion (vane) attached to either side of the rachis. Notice that the vane is made up of a number of elastic slender laminæ on barbs. These are held together by means of the minute processes (barbules) which form a fringe on each side of the barb.

(4) The minute supplementary feather (after-shaft) which is attached to the ventral surface of the central axis just where the rachis joins the quill.

Make a sketch of one of the quill feathers, showing the various structures described above.

Now pluck all the small feathers off the bird and observe the comparative length of the neck; the prominence of the central ridge of the breast-bone (sternum), to which the chief muscles of flight are attached, the situation of the transverse slit-like cloacal aperture below the tail, where the digestive, excretory and reproductive organs all terminate, and the aperture of the uropygial gland, which lies in the middle line above the tail.

Dissect off the skin from one side of the breast, so as to expose the first or great pectoral muscle, which is the chief agent in depressing the wing in flight. It is attached at one end along the sternum and collar-bone (clavicle), whence its fibres converge to a tendon which is attached to the humerus. Cut through the muscle carefully about the middle of its length, taking care not to cut into the second pectoral muscle, which lies below. Turn back the two flaps so as to see how the muscle is attached, and clean the tendon by which it is connected with the humerus. Now dissect away all the basal portion of the muscle, so as to expose the collar-bone (clavicle) on one side. Continue the dissection on the other side, so as to see how the clavicles are united together in the middle line to form what is popularly known as the merry-thought.

On the side from which the great pectoral muscle has been completely removed, you will now have exposed the second pectoral muscle. The second pectoral is very similar both in its method of attachment and in its shape to the great pectoral, but its tendon is very much longer. It passes through a pulley-like arrangement to its place of attachment at the back of the shoulder-joint. It thus serves to elevate the wing in flight, as can be readily seen on pulling the muscle with your forceps. You will now be able, without much difficulty, to recognise the chief bones of the pectoral girdle, which is a very characteristic feature in the anatomy of birds. It consists of the following:—

(1) The scapula, which is a long flattened bone lying behind the shoulder above the ribs.

(2) The coracord, which is a stout straight bone running from the shoulder to the front end of the sternum.

(3) The clavicle, which is ossified to the corresponding bone on the other side of the body, the two together making up the merry-thought, which we have already examined.

You will also be able to make out the nature of the ventral aspect of the sternum and ribs.

Now dissect off the whole of the sternum, cutting through the ribs on either side and the coracoids and clavicles in front with a pair of scissors, taking great care not to injure any of the organs which lie below. You will now have exposed the following organs:—

(1) The ends of the posterior abdominal air-sacs. These are sacs of transparent membrane to be found on either side immediately under the posterior ribs; they should be looked for

before any of the other organs are interfered with, as they collapse when cut into and are then difficult to recognise. The air-sacs are numerous and lie in different parts of the body cavity and neck; they are difficult, however, to dissect out completely. Filled, as they are, with warm air, they serve to increase the buoyancy of the body in the atmosphere. When uninjured, they can be inflated by means of a blow-pipe inserted in the trachea.

(2) In the middle line in front you will be able to see the base of the trachea where it divides into two branches (bronchi) which communicate with the lungs. You can at once recognise the trachea by the close-set cartilaginous rings by which it is supported. Just above the insertion of the bronchi is the organ of voice (syrinx). This is clothed with muscles which appear like a swelling in the trachea and should be examined carefully, as the manner of their insertion into the ends of the bronchial semi-rings is a character used in classifying the group (Passeres) to which the sparrow belongs.

(3) Immediately below the syrinx, in the middle line, is the large heart with its numerous vessels. The heart in birds is divided into four chambers, *viz.*, a right and left auricle and corresponding ventricles. The ventricles occupy the thick muscular apex of the heart and the partition between them cannot be recognised without further dissection. The right and left auricles can be easily distinguished as thin-walled vessels filled with blood, situated on either side in front of the thick-walled ventricles. Opening into the right auricle you will be able to see the base of the large thin-walled vein (right anterior vena cava) which returns the blood from the jugular, pectoral and brachial veins. In the middle line, arising between the auricles, you will see the two white thick-walled innominate arteries, which the aorta gives off almost immediately after leaving the right ventricle. By moving the tissues a little, but without further dissection, you will be able to trace where the innominate artery on either sides breaks up into the carotid artery, which supplies the side of the head, and the sub-clavian artery, which supplies the wing and breast.

(4) Behind the heart lie the large, dark-coloured lobes of the liver.

At this stage in the dissection the bird should be pinned out under water, so as to float the organs that are exposed. Dissect off the soft wall of the abdomen, which lies between the sternum and the cloacal aperture and clear away the membranes by which the various organs are attached to each other.

The vessels of the heart can now be examined more in detail. After clearing the membrane (pericardial membrane) from the heart, it can be turned up from behind so as to expose the posterior vena cava, which returns the blood from all the posterior parts of the body into the right auricle. This vessel passes through the large right lobe of the liver, where it receives the hepatic veins. Behind the liver it divides into the right and left iliac veins. These again split into femoral, hypogastric and renal veins, which return the blood from the leg, the lower part of the intestine and the kidneys respectively. After making out the posterior vena cava and tracing it into the lobe of the liver, examine the posterior part of the left auricle for the pulmonary vein, which returns the blood from the lungs. Then clean the bases of the vessels in front of the heart, and trace back the two innominate arteries to where these arise in the form of the aorta from the left ventricle. This vessel, after giving off the innominate arteries, continues its course downwards round the right bronchial tube as the dorsal aorta, which supplies blood to the digestive tract and all the posterior part of the body. Search also for the pulmonary artery, which arises from the base of the right ventricle close to, but below, the aorta and supplies blood to the lungs. Then cut across the ventricles and see the partition which separates the two chambers. Trace the trachea up to the glottis and then cut it away altogether. Cut away the heart. Unravel the digestive tract, noticing the veins in the mesentery, which collect the blood from the digestive tract and converge to form the portal vein, which carries the blood into the liver. Make out the following parts in the digestive tract :—

(1) The wide crop formed by the dilation of the posterior part of the œsophagus. Cut across the crop to see its thin walls.

(2) The slightly expanded stomach. Cut it across to see its walls, which are thick and glandular as compared with the thin walls of the crop. Notice the small red glandular organ (spleen) attached to the side of the stomach.

(3) The thick hard muscular gizzard, in which the food is crushed, after being acted upon by the gastric juices secreted in the stomach. Notice how the stomach and duodenum enter and leave the gizzard close to each other. Cut the gizzard across to see its thick muscular walls.

(4) The U-shaped loop of the intestine (duodenum) which lies immediately behind the gizzard. The centre of the U-

shaped loop is taken up by an elongated pinkish gland (pancreas), which pours its secretion into the digestive tract. The duodenum also receives the bile secreted by the liver.

(5) Behind the duodenum the intestine continues its course as a long narrow much convoluted tube to the cloacal aperture.

(6) About half an inch above the cloacal aperture you will see a little white ear-like appendage on either side of the tube. These are cæca, which in some birds are very largely developed. They correspond to the vermiform appendage which marks the commencement of the large intestine in mammals.

Cut away the whole of the digestive tract, including the liver, and wash out the body cavity. You will now be able to see —

(1) The yellowish spongy lungs which fill the front portion on either side of the back of the cavity. A few cuts with the knife on either side will enable the lungs to be removed bodily without injury to the nerves behind. Notice how closely the lungs are appressed into the spaces between the ribs.

(2) Posterior to the lungs lie the dark-coloured excretory glands (kidneys). Their ducts pass down on either side to the urogenital pouch, where the digestive tract, the excretory and the reproductive organs, all unite before passing out at the cloacal aperture.

(3) At the front end of the kidneys are a pair of somewhat indistinct small yellowish bodies known as the supra-renal bodies. They must not be confounded with the testicles.

(4) Above the kidneys in the male are a pair of round whitish testicles which vary a good deal in size at different times of the year. In the cold weather in Calcutta they are conspicuous organs as big as peas. In the female the left ovary alone is developed ; it is a thin membranous sac lying in front of the left kidney. It is usually dotted with eggs, which are often very minute.

Cut away the reproductive organs and kidneys. You will now have the ribs and vertebral column exposed. Notice the white (thoracic spinal) nerves which pass out from the spinal column between each rib. Notice also the delicate longitudinal nerve cords on either side of the spinal column, with swellings at intervals. These connect the roots of the spinal nerves and are known as the sympathetic nerve system. By clearing away the tissues at the base of the neck you will be able to see the net-work of large nerves known as the brachial plexus, which supply the shoulder and wings. The brachial plexus is formed by the junction of the posterior cervical and the anterior thoracic spinal nerves. By clearing away the tissues in the region of the kidneys you will be able to see the lumbar plexus and the sciatic plexus, which arise from the lumbar and sacral vertebræ respectively and supply nerves to the leg.

Separate the head and neck from the rest of the body by cutting through the base of the neck. Skin and then cut away the back of the head carefully with a knife, so as not to injure the brain, which lies immediately beneath the bone. Pick away the sides of the skull so as to expose the whole of the brain, and cut away the dorsal arches of the vertebral column with scissors. Lift up the brain from behind with as much of the spinal cord attached as possible. Cut through the numerous nerves which pass out through the skull below, so as to remove it completely. Examine it under water with a lens and make out the following parts:—

(1) The spinal cord, which fills the canal above the back-bone and passes up into the skull by the large hole (foramen magnum) behind.

(2) The medulla oblongata or expanded part of the spinal cord after it enters the skull.

(3) The median striated lobe (cerebellum), which overhangs the medulla oblongata above.

(4) The large smooth cerebral hemispheres, which occupy almost all the upper part of the cavity of the skull and meet each other down the middle in a straight suture.

(5) The olfactory lobes, which are two little protuberances in which the cerebral hemispheres terminate in front. From these the olfactory or first nerves are given off to the nasal sacs.

(6) Between the cerebral hemispheres and the cerebellum lies a small lobe known as the pineal body; this, however, is so much overhung by the cerebral hemispheres that you will probably be unable to make it out.

Now turn the brain over in the water and on its ventral surface observe—

(1) The lower surface of the cerebral hemispheres in front.

(2) The large rounded optic lobes which lie one on either side below and behind the cerebral hemispheres. They are connected with each other by the large optic tracts which pass inwards from each optic lobe and cross in the middle line before giving origin to the enormous optic, or second nerves which supply the eyes.

(3) The infundibulum, which lies in the middle line immediately behind the junction of the optic tracts. It is connected with the pituitary body, which lies below.

(4) Immediately behind the infundibulum, on either side, you will be able to make out the pair of third nerves, which have their origin close to the middle line.

(5) The fourth pair of nerves pass down on either side from their origin on the dorsal surface and can be recognised lying between the optic lobe and the medulla oblongata.

(6) Behind the nerves above noticed, there can be made out, under favourable circumstances, the pairs Nos. 5 to 12, which all have their origin in connection with the ventral portion of the medulla oblongata. You are not likely, however, to be able to make them out satisfactorily without more elaborate dissection than is here intended.

Cut off one of the legs with scissors across the middle of the tibio-tarsus. Dissect off the skin down to the toes and notice the tendons which lie at the back of the metatarsus. By pulling various parts of the muscles that have been cut through with the tibio-tarsus, the students will be able, without difficulty, to separate out two tendons which serve to flex the hallux and the three opposed toes respectively. On stripping off the sheaths of these tendons they will be able to see that the tendons pass down together behind the tarso-metatarsus, but are quite distinct from each other, so that the flexing of the hallux is independent of the flexing of the toes. The tendon which flexes the hallux is known as the flexor longus hallucis, while the tendon which flexes the three opposed toes is the flexor perforans digitorum. Their arrangement is an important characteristic of the group of birds to which the sparrow belongs.

Take the remains of the skull, from which the brain has been extracted. Cut off the lower jaw, strip off the horny beak from the upper jaw, and pick away the flesh and membrane from the roof of the mouth with forceps, so as to expose the bones of the palate. Notice that the central bone (vomer) which lies in the roof of the mouth, just below the anterior end of the partition which separates the eyes, and which bounds the cavity of the nostrils behind, is broad and blunt and disconnected from the little bones (maxillo-palatines) which lie on either side of it. This arrangement is one of the characteristics of the order of birds to which the sparrow belongs. It is difficult however to make out satisfactorily in the sparrow.

DISSECTION OF THE PALM SQUIRREL.[1]

To make out the essential features of Mammalian structure a palm squirrel (*Sciurus palmarum*) may conveniently be taken. Palm squirrels are excessively common in Dehra, where they are known as *gilehri*. They are best trapped or netted. They can be readily killed by putting them into any closed receptacle and pouring in chloroform.

[1] It may be necessary to curtail or omit this dissection for want of time.

Examine a specimen, noticing that the whole body is covered with hair (as opposed to the feathers of the bird and the scales of the reptiles). In the head make out the mouth, nostrils, eyes and ears. Notice that the body consists of an interior portion, or thorax, protected by ribs and a posterior portion, or abdomen, which is not so protected.

In the fore-limb make out—(1) the basal joint (arm), which is supported by the humerus bone ; (2) the second joint (fore—arm), which is supported by the radius and ulna bones ; (3) the wrist, supported by a number of small bones (carpals) ; and (4) the hand, which consists of four digits, each terminating in a claw, the thumb being rudimentary.

In the hind limb make out—(1) the basal joint (thigh), supported by the femur, which is articulated to the pelvic girdle of bones that can be felt through the skin in the lower portion of the back ; (2) the second joint (crus), supported by the tibia and fibula bones ; (3) the foot, which comprises a basal portion supported by a number of tarsal bones followed by five digits, each armed with a claw.

Feel the skull and vertebral column (back-bone) through the skin, noticing how the latter is produced into the tail. In the female notice the teats, which communicate with the mammary glands. In the male notice the large scrotal pouch, which in the breeding season contains the testes, which are of large size. In the case of both males and females notice that the anal and urogenital openings are quite distinct from each other.

Pin the animal on its back to the weighted bark, the pins being inserted into the bases of the legs. Immerse it in water. Slit open the skin down the entire ventral surface in the middle line. Dissect away and pin back the flap of skin on either side. In the female notice the extensive mammary glands which lie immediately beneath the skin on the ventral surface of the abdomen.

In connection with the neck, the large jugular vein, which will be found on either side, usually gorged with blood, should be made out (it conveys blood to the anterior vena cava, which will be noticed in connection with the heart) ; also three pairs of pale-pink salivary glands. These salivary glands are—

(1) The parotid gland, which is situated on the side of the head below the ear ; it pours its secretion into the mouth by a duct which opens in the cheek of the upper jaw.

(2) The sub-maxillary gland, situated in the neck near the middle line, at the posterior angle of the mandible. It pours its secretion into the mouth by a duct which opens below the tongue.

(3) The sublingual gland, which is much smaller and lies just above the sub-maxillary gland, its secretion being poured into the mouth by a duct which opens below the tongue.

The students need not attempt to make out the ducts of these salivary glands.

Dissect away the pectoral muscles, so as to expose the ribs. Make out—

(1) The sternum or breast-bone, to which the ribs are attached in the middle line in front.

(2) The slender collar-bone (clavicle), which is attached at one end to the anterior end of the sternum and at the other end to a process of the flat shoulder-blade (scapula), the humerus bone being articulated between them. Dissect away the muscles of the shoulder, so as to make out the relative positions of these three bones.

(3) The ribs, which pass round the thorax from the vertebral column.

Slit open the soft abdominal cavity behind the ribs, taking care not to injure the viscera, which lie below. Pin back the muscular wall and make out the following organs:—

(1) The liver, which is a large dark-reddish gland lying beneath the posterior end of the ribs. Notice the little green gall bladder in a notch of the outer lobe on the right side.

(2) The stomach, which is a large sac-like expansion of the digestive tract, lying immediately behind the liver.

(3) The winding intestine, which takes up the greater part of the rest of the abdominal cavity and terminates in the anus behind.

(4) The thin-walled urinary bladder, which lies at the posterior end of the abdominal cavity and communicates with the urogenital opening in front.

Turn the viscera over without cutting anything. Notice the partition in front (diaphragm) which separates the abdominal cavity from the thoracic cavity, and make out the following organs:—

(1) The spleen, which is an elongated glandular organ lying in the left of the body behind the stomach.

(2) The kidneys, which are a pair of bean-shaped bodies lying in the back of the abdominal cavity. Trace the pair of slender yellowish tubes (ureters) by which the secretion of the kidneys passes into the bladder. Notice the little rounded organ (adrenal body) at the anterior end of each kidney.

(3) The fold of transparent membrane (mesentery) which supports the digestive tract and contains numerous blood-vessels. Endeavour to make out the omentum, which is a fold behind the stomach sometimes loaded with fat.

(4) If the specimen is a female notice the thick-walled organs (uteri) which bifurcate behind the bladder and vary very much at different times of the year. If the specimen is a male, notice the paired testes, which can sometimes be pulled back into the abdominal cavity from the scrotal sacs. Notice also the bilobed uterus masculinus behind the bladder, which receives the vasa deferentia from the testes and itself communicates with the urogenital canal.

Cut through the ribs on either side carefully with scissors and dissect off the breast-bone (sternum), together with the whole of the front wall of the thoracic cavity, so as to expose the organs which lie below. Observe that the thoracic cavity is completely shut off from the abdominal cavity, and is lined with glistening membrane. In the middle lies the large rounded heart enclosed in transparent membrane (pericardium), while on either side are the spongy lungs.

Clear away the muscles in front, so as to make out the wind-pipe (trachea), through which air passes into the lungs. It lies in the middle line of the neck and can be immediately recognised by the thickened cartilaginous rings by which it is supported.

Clear away the glandular matter (thymus) from in front of the thoracic cavity, dissect off the membrane in which the heart is enveloped, and clean the vessels at the base of the heart. By pressing the heart and the lungs apart a white nerve chord (phrenic nerve) can be seen on either side. These nerves are branches of the fourth cervical nerves and supply the muscular partition (diaphragm) which separates the thoracic and abdominal cavities.

Examine the heart and make out—

(1) The thick muscular ventricles which occupy all the posterior portion of the organ, the partition between the right and left halves not being very plainly visible externally.

(2) The right and left auricles, which are two thin-walled chambers lying in front of the right and left ventricles. To see the auricles plainly the heart should be turned over a little first to one side and then to the other. The right auricle receives the blood from all parts of the body except the lungs, and passes it on into the right ventricle. The right ventricle drives the blood to the lungs, whence it returns to the left auricle, which passes it on into the left ventricle. The left ventricle drives the blood to all parts of the body, whence it finds its way back to the right auricle, thus completing the circle.

The chief vessels in connection with the heart are—

(1) The right and left vessels (anterior vena cava) by which the blood returns to the right auricle from the anterior portion

of the body. These large veins can be made out without much difficulty, passing down along the inner side of each lung. The right anterior vena cava passes directly into the right auricle in front, while the left anterior vena cava has to cross the heart behind and enters the right auricle on the left side.

(2) The posterior vena cava. This large vein returns the blood from all the posterior parts of the body. It can readily be seen, on turning up the base of the heart, where it enters the posterior end of the right auricle. It traverses the liver and is again visible in the posterior part of the abdominal cavity on the right side. Its chief branches are the hepatic veins, which return the blood from the liver, the renal veins, which return the blood from the kidneys and the iliac and lumbar veins, which return the blood from the posterior extremities.

(3) The pulmonary artery. This is a somewhat smaller thick-walled vessel. It arises from the anterior end of the right ventricle and is the channel through which the blood is driven into the lungs. On pulling back the heart with the finger and thumb it can easily be seen where it leaves the right ventricle, between the two auricles, and arches over behind the left auricle before branching to the lungs.

(4) The pulmonary veins. These veins return the blood from the lungs to the left auricle. Turn up the heart from behind and make out the two branches from the lungs, which unite before passing with the dorsal portion of the left auricle.

(5) The aorta. This thick-walled vessel arises from the left ventricle just behind the origin of the pulmonary artery. It is the channel through which the blood is pumped by the left ventricle. Trace it forwards as far as possible and see where it arches over to the left, after giving off the large vessels (innominate and left sub-clavian arteries) which supply the head and fore-limb. By pulling the heart back with the finger and thumb the three branches (left carotid, right sub-clavian and right carotid) into which the innominate artery breaks up can easily be made out. Notice the ligamentous band (ductus arteriosus) which connects the arch of the aorta with the pulmonary artery, and which represents a channel of communication between these two vessels through which blood passes during embryonic life. On pulling back the left lung the continuation of the aorta can be seen where it passes down the dorsal portion

of the chamber to supply blood to the posterior organs of the body.

Now turn to the abdominal cavity and endeavour to make out the vessels which unite to form the portal vein, which returns the blood from the digestive tract and spleen to the liver, where it passes through a series of capillaries before joining the posterior vena cava. The chief of these vessels are—(1) the lineogastric vein, which returns the blood from the stomach and spleen; (2) the duodenal vein, which returns the blood from the duodenum; and (3) the anterior and posterior mesenteric veins, which return the blood from the whole of the intestine and rectum. These veins are best seen in a fresh specimen; the important point to remember about them is that the blood from the whole length of the digestive tract returns to the liver.

Free the coils of the digestive tract by cutting through the membranous mesentery by which it is supported. Straighten out the intestine. Trace up the narrow tube (œsophagus) which passes through the diaphragm and serves to convey the food from the mouth into the stomach. Observe that the whole digestive tract is merely a winding tube which connects the mouth in front with the anus behind, and notice the following portions of it:—

(1) The narrow œsophagus.

(2) The expanded stomach.

(3) The narrow portion (duodenum) which leaves the right side of the stomach and receives in front the bile duct from the liver, and behind the duct from the pancreas, which is a gland situated in the loop formed by the duodenum and is very small in the palm squirrel.

(4) The small intestine, which is the narrow winding portion of the tube between the duodenum and a large blind sack-like outgrowth known as the cœcum.

(5) The remainder of the tract consist of the colon and rectum, where the tube is somewhat wider. The rectum opens at the anus behind.

Cut away the whole of the digestive tract, also the liver, diaphragm, heart, and lungs, with scissors. The dorsal aorta can now be readily traced down the back, immediately above the vertebral column on the left side. Dissect it out as far as possible, noticing the branches that it gives off in its course.

Dissect away the muscles of the shoulder and hip, so as to see the large nerves (brachial plexus) which supply the fore-limb, also the large nerves (lumbo sacral plexus) which supply the hind-limb. These nerve plexi are formed in each case by the union of several nerves which issue from the spinal cord, but it will not be necessary to trace them to their origins.

Force open the mouth and examine the pair of large chisel-shaped teeth (incisors) both in the upper and lower jaw. These teeth continue to grow throughout life, and are characteristic of the order (Rodentia) to which the palm squirrel belongs.

Cut away the soft skin of the roof of the mouth, and make out where the cavity of the nostrils passes out behind just opposite to where the trachea begins, so that the air is able to pass in breathing directly into the lungs without entering the mouth. Eustachian tubes connect the cavities of the ears with the posterior nasal chamber, but the students will probably be unable to make them out.

Turn the specimen over and pin it out into the weighted bark with the back uppermost. Take off the skin, and dissect away the muscles from the back of the neck. Insert the point of the scissors into the large opening (foramen magnum) of the skull behind, and carefully dissect off the back and sides of the skull, taking care not to injure the soft brain that it contains. Continue the dissection down the back, cutting through the spinal arches so as to expose the spinal cord. Hardening the brain by placing the specimen in strong alcohol for a few days after the roof of the skull has been removed facilitates subsequent dissection, but the more important features can be made out in the fresh state.

Notice that the skull is lined with tough membrane (dura mater), which will probably be torn off in removing the bone. Beneath this is a much thinner membrane (pia mater), which closely invests the brain and contains the superficial blood-vessels. Pick the brain as clean as possible with forceps and make out the following parts, beginning from behind :—

(1) The medulla oblongata, which is the expanded end of the spinal cord where it passes into the skull.

(2) The cerebellum, which is a much foliated mass lying above the medulla oblongata.

(3) The cerebral hemispheres which are two smooth elongated masses in front of the cerebellum. They take up the greater part of the upper portion of the cavity of the skull.

(4) Lying in the angle between the cerebral hemispheres and the cerebellum is the small pineal body, which is very liable to get torn away in removing the roof of the skull. It represents the rudiment of what was once a median eye.

(5) The olfactory lobes, which are two small rounded masses lying in front of the cerebral hemispheres. They give origin to the olfactory nerves.

Roy. 8vo. Cloth. pp. v, 406. 21s.

A

HISTORY OF THE DECCAN

WITH NUMEROUS ILLUSTRATIONS, PLATES, PORTRAITS, MAPS, AND PLANS.

VOL. I.

From the Origin of the Bahmanee Kings till the beginning of the Eighteenth Century.

BY

J. D. B. GRIBBLE.

"In a style easy and pleasant the author tells the story of the Mohammedan occupation of the Deccan . . . . the general style of the book and the admirable photographs and drawings with which it is enriched leave nothing to be desired." —*Athenæum.*

"Mr. J. D. B. GRIBBLE has accomplished a difficult task. He has constructed from original materials a continuous narrative of one of the most confused periods of Indian history. He has also presented it with a lucidity of style which will go far to render it acceptable to the reading public . . . . The book is illustrated by a number of interesting reproductions of scenery and architecture in Southern India. These and the maps, plans, and clear genealogical tables reflect credit both upon the author and the publisher."—*Times.*

"Mr. GRIBBLE has brought great industry and knowledge of the country to this compilation . . . . The work is of some historical importance."—*Saturday Review.*

LONDON: LUZAC & CO.,

Publishers to the India Office,

46, GREAT RUSSELL STREET, OPPOSITE THE BRITISH MUSEUM.

(6) By turning up the cerebral hemispheres behind, the posterior portion of the rounded optic lobes, which lie below, can be made out.

Cut through the spinal cord about half an inch behind the cerebellum. Take hold of it with forceps and gently lift it up so as to raise the posterior portion of the brain from the floor of the skull. Notice the nerves which arise from the sides and ventral surface of the medulla oblongata. These nerves are eight in number, being the fifth to the twelfth cranial nerves. Most of them have numerous roots. The chief of them are the hypoglossal, which supplies branches to the tongue, the pneumogastic, which supply branches to the stomach, lungs, throat, and heart, the auditory nerve, which supplies the ear, and the trigeminal nerve, which supplies the front of the head. The student will not be able to distinguish between these nerves without more elaborate dissection than is here intended, but he should notice the roots he is able to make out.

Now lift the brain completely out of the skull, cutting through the nerves with the point of a knife. Examine the lower surface under water, and make out :—

(1) The two large temporal lobes of the cerebral hemispheres, which stand out as rounded protuberances on either side, about the middle of the length of the brain.

(2) Immediately between the large temporal lobes is a median rounded elevation (infundibulum) to which the small rounded pituitary body is attached. The pituitary body is usually torn off in extracting the brain.

(3) In front of the infundibulum may be seen the optic chiasma, where the large optic tracts cross each other before passing into the eyes. Trace back the optic tracts, by pushing aside the temporal lobes, to the large optic lobes above.

The sympathetic nerve system in the palm squirrel is difficult to recognise. It consists of a series of delicate nerve-cords and ganglia lying on either side of the dorsal aorta, but the students are not likely to be able to make them out.

NOTE ON THE PRESERVATION OF ZOOLOGICAL SPECIMENS.

If the animal is a small one, the simplest plan is to cut one or two slits in it, so as to enable the alcohol to penetrate thoroughly, and then to drop it into a bottle of spirits of wine. The spirit should be changed two or three times, as it gets weak by absorbing moisture out of the specimen. All animals, including insects, can be preserved in spirit in this way. The method, however, is only practicable with small animals, and even in their case the alcohol damages the colours and renders butterflies and moths almost unrecognisable. It is, however, the best way to preserve all caterpillars, grubs and other soft-bodied insects, also small snakes, lizards, fishes and other small vertebrates. To prevent the shrivelling which is liable to take place when the specimen is dropped at once into strong alcohol, it is a good thing to begin by putting it for a few hours into weak spirit, which should afterwards be changed for the very strongest available.

All hard-bodied insects, including moths, are best killed in a killing bottle and then wrapped up in thin paper to dry, after which they can be preserved for any length of time, provided they are kept in a tight-fitting box with a little camphor, or naphthaline, to prevent their being attacked by mites and beetles. In the case of insects with large thick bodies, before wrapping them up in paper, some of the viscera should be removed through a slit made with a pair of scissors on the lower surface of the abdomen. The abdomen should then be stuffed with a little cotton wool on which a few drops of carbolic acid have been poured. For most butterflies no killing bottle is required, as they can easily be killed by pinching the thorax between the finger and thumb.

A killing bottle is made by putting a few lumps, each as big as the top of one's thumb, of Cyanide of Potassium (which should be handled carefully, as it is extremely poisonous) at the bottom of a bottle and then pouring in sufficient plaster of Paris to cover the lumps. The plaster of Paris is first mixed with enough water to make it of the consistency of cream. The plaster of Paris soon sets into a firm mass at the bottom of the bottle. When it is dry, the bottle should be tightly corked up to prevent the escape of the vapour of the cyanide. Any insect can be killed in a few minutes by dropping it into such a bottle and closing up the cork.

For vertebrata, which are too large to preserve entire in alcohol, the best thing is to preserve the skins. The skin should be removed by slitting it open down the ventral surface of the body (or, in the case of fishes, down one side), and then gradually dissecting it away from the body and limbs, using plenty of wood-ashes to prevent soiling. The skull should generally be left in the skin, but it should be cleaned as much as possible and the brain should be picked out through the large opening (foramen magnum) behind. To do this the skin should be turned inside out over the head. The eyes also should be carefully extracted. In the case of small vertebrates the leg bones (in birds also the wing bones) should be left in the skin, but they should be cleaned as much as possible. In the case of fish it is a good thing to paste the whole animal up in thin paper before skinning it, as this prevents the scales coming off in the operation. The paper can easily be washed off afterwards. The eyes in birds are very large, and the best way to remove them is by opening the mouth and cutting them out with a pair of scissors through the palate. After the skin has been cleaned as much as possible of fat and flesh, it should be well rubbed on the inside with arsenical soap to poison it. After this has been done, the skin can be preserved indefinitely, provided it is thoroughly dried. In the case of fishes, the skin shrinks so much in drying that it is best to stuff it tightly with cotton wool and then to sew up the slit in the side before drying. In the case of birds all that is necessary is to stuff the eye-sockets and throat and put a little cotton wool into the neck and body to preserve the lie of the feathers. A small piece of stick cut a little longer than the combined length of the body and neck and inserted, one end into the back of the skull and the other into the base of the tail, helps to keep the skin in a natural position in drying.

The proportions of the various ingredients for making arsenical soap are given in Hume's *Vade Mecum* as follows:—

One part of camphor.
One part of spirits of turpentine.
Eight parts of soap.
Eight parts of white sublimated arsenic.

The soap should be first melted over the fire with a little water to prevent burning. The arsenic should be mixed in and well stirred while the soap is hot. (*N.B.*—The arsenic must be handled carefully, as it is very poisonous.) The camphor and turpentine, which should previously have been mixed together, are to be woked in as soon as the soap is cool enough for the hand to rest in it. It is most important that the ingredients should be thoroughly mixed together.

In preparing skeletons or skulls the bones should be picked as clean as possible by cutting the flesh and skin off them, care being taken not to lose any of the small bones (especially the tongue bones, the knee-cap and the small bones of the wrist). The bones should then be dried in the sun, and packed up in saw-dust. Bones should never be boiled to remove the flesh, but when wanted for setting up, the sun-dried skeleton should be macerated in cold water until the flesh falls off, when the skeleton can easily be cleaned.

EXPLANATIONS OF TECHNICAL TERMS.

Amœboid.—Having movements like the Amœba.
Annulate.—Marked with rings.
Antenna.—A jointed appendage attached to the head in Arthropoda.

Aortic arch.—A large artery which conveys the blood from the heart towards the back in Vertebrata.

Asexual.—Without the intervention of the male.

Auditory.—Connected with hearing.

Auricle.—A thin walled chamber of the heart in Vertebrata.

Barbels.—Slender feeler like organs situated about the mouth in many fishes.

Bilaterally symmetrical.—With body arranged in such a manner around a central axis that the two sides are similar to each other, but the back (dorsal surface) is different from the front (ventral surface).

Branchiæ.—Leaf-like respiratory organs adapted for breathing oxygen dissolved in water.

Canines.—A pair of teeth situated one on each side of the mouth in Mammalia.

Carabid.—Belonging to the group Carabidæ.

Caudal.—Relating to the tail.

Cell.—The ultimate living unit of which the body is built up. In its simplest form it consists of a little speck of protoplasm containing a nucleus.

Cephalic.—Connected with the head.

Cephalothorax.—A divisiön of the body composed of the head and thorax combined. It occurs in spiders and many crustacea.

Chitin.—The substance of which the armour plates in insects are composed.

Chitinous.—Formed of chitin.

Chrysalis.—The name popularly given to the pupa in the group Lepidoptera on account of the golden coloration noticeable in certain species.

Cilia.—Little vibratile hair-like processes attached to cells.

Ciliated.—Furnished with cilia.

Cœlomate.—Provided with a body cavity in which the digestive tract is suspended.

Coleopterous.—Belonging to the group Coleoptera.

Dermal.—Connected with the skin.

Digestive tract.—The organs of the body which serve to take up nutriment direct from the food.

Digitigrade.—Walking upon the tips of the toes.

Dipterous.—Belonging to the group Diptera.

Dorsal.—Relating to the back or upper surface of the body.

Elytra.—The stiff horny front wings in beetles.

Epidermis.—One of the layers of the skin in Vertebrata.

Excretory.—Connected with the casting off of waste products.

Femur.—One of the joints of the leg.

Filiform.—Resembling a thread.

Fœtus.—The young animal before it quits the body of its mother.

Heteromerous.—Belonging to the group Heteromera—with tarsi four jointed on the hind legs and five jointed on the first and second pairs of legs.

Hymenopterous.—Belonging to the group Hymenoptera.

Imago.—The mature stage in the life history of an animal.

Incisors.—A group of front teeth in Mammalia.

Inguinal.—Connected with the groin.

Integument.—The outer layer of the body.

Labial.—Connected with the labium.

Labium.—The lower lip in insects.

Labrum.—The front or upper lip in insects.

Larva.—The imperfectly developed stage through which many animals pass after leaving the egg.

Mammary glands.—Glands characteristic of the Mammalia which secrete the milk with which the offspring is nourished for a time after birth.

Mandible.—One of the front pair of jaws in Arthropoda; also the chief bone of the lower jaw in Vertebrata.

Mantle.—In the Mollusca, a fold of the integument which generally secretes a shell.

Maxilla.—One of the second pair of jaws in insects; also a bone in the upper part of the mouth in Vertebrata.

Maxillary.—Connected with the maxilla.

Mesothorax.—The middle segment of the thorax in insects.

Metathorax.—The posterior segment of the thorax in insects.

Migratory.—Prone to wander long distances.

Molars.—Certain back teeth in the Mammalia.

Mucous membrane.—The slimy membrane which lines the digestive tract in the higher animals.

Multicellular.—Formed of many cells.

Nucleus.—A dense structure which is to be found in every cell.

Occipital condyle.—A bony process situated at the base of the skull in Vertebrata.

Ocelli.—Minute eyes of very simple structure possessed by many insects.

Organ.—A portion of the body differentiated for the performance of a particular function.

Organism.—That which is made up of organs—an animal.

Orthopterous.—Belonging to the group Orthoptera.

Palp.—A jointed organ attached to some of the appendages of the mouth in Arthopoda.

Pectoral.—Relating to the breast.

Pectinated.—With processes resembling the teeth of a comb.

Peduncle.—A stalk such as that by which the abdomen is connected with the thorax in many Hymenoptera.

Pelvic.—Relating to the Pelvis.

Pentamerous.—Belonging to the group Pentamera, with five-jointed tarsi.

Plantigrade.—Walking upon the sole of the foot.

Plumose.—Furnished with feathery hairs.

Premolars.—Certain teeth in the Mammalia situated between the molars and the canines.

Prolegs.—Fleshy protuberances which serve as additional legs in the larvæ of some insects.

Prothorax.—The anterior segment of the thorax in insects.

Protoplasm.—A jelly-like substance which is an essential compenent of every cell.

Pupa.—The resting stage which intervenes in the life history of many insects between the larval and imago stages.

Pupal.—Relating to the pupa.

Quadrate bone.—A bone situated in the side of the head in Vertebrata.

Radially symmetrical.—With body arranged symmetrically around a central axis, the ventral surface not being differentiated from the dorsal.

Ruminant.—That chews the cud.

Scutellum.—A triangular plate situated in the middle of the back at the base of the front wings in Coleoptera and many Rhynchota.

Segments.—The joints of which the body in Arthopoda is made up.

Segmentation.—Division into segments.

Segmented.—Composed of segments.

Sessile.—Attached direct without the intervention of a stalk.

Stigmata.—The external openings of the breathing tubes in insects.

Stridulation.—Sound made by many Arthropoda. It is usually produced by the rubbing of one rough surface against another.

Tarsus.—A joint, or series of joints, in the leg.

Tetramerous.—Belonging to the group Tetramera, with four-jointed tarsi.

Thoracic.—Belonging to the thorax.

Thorax.—One of the divisions of the body in Arthropoda and Vertebrata.

Tibia.—In insects, one of the joints of the leg: in Vertebrata, one of the leg bones.

Tissues.—The living materials of which the body is made up.

Tracheæ.—The breathing tubes in insects.

Trimerous.—Belonging to the group Trimera with three-jointed tarsi.

Unsegmented.—Not composed of segments.

Urogenital.—Connected with the urinary and reproductive organs.

Urticating.—Capable of inflicting a sting.

Uterus.—That portion of the reproductive organs in the female where the young animal is elaborated before quitting the body of its mother.

Ventral.—Relating to the front or lower surface of the body.

Ventricle.—A thick walled chamber of the heart in Vertebrata, which receives the blood from one or both of the auricles.